THE AI READY™ EXECUTIVE SERIES · BOOK ONE

AI READY
IN 90 DAYS

A Leadership Roadmap for Sustainable AI Adoption

VALETTA TAYLOR

Published by The Harmony Collection by Valetta Taylor, LLC · Houston, Texas

ISBN 979-8-9965901-2-4 (paperback) · First Edition

ecfsystem.com/aiready

For every executive I've ever worked with who started a project without the information in this book.

Whether it's AI, a new system, a new process, an integration, or a merger — all of it requires readiness: clarity about where you start, what you measure, what success looks like, and where you finish. If anything before the finish is unclear, finishing becomes difficult. And if the finish is the only thing you measure, you've failed before you started.

My dad, a former Marine, carried a saying through my entire career — he credits the Corps: the Five P's. Proper Planning Prevents Piss-Poor Results. Private sector, public, government — it holds everywhere. Planning, understanding, assessing, evaluating, and above all acting, so that you and your organization are ready: that is what this book is about.

It's written from the seat of a transformation executive — the person challenged to implement, fix, and orchestrate these things every single day. The truth is, we can't do our best work without you: your clarity, your sponsorship, your resources, your communication, your people.

This one's for you.

Contents

Prologue: We Celebrated the Wrong Victory....1
Introduction: The Technology Was Never the Problem....5

PART ONE — AWARENESS....10

1. Why AI, Why Now?....11
2. Why AI Isn't Delivering What Leaders Expected....17
3. The Fear Nobody Names....24
4. The People AND the Machine....32
5. Shadow AI Is Already in Your Building....39
6. Communicate Why Before What....46

PART TWO — DIAGNOSE....54

7. You're Solving the Wrong Problem....55
8. You're Measuring the Wrong Things....62
9. Do You Have the Right Leaders?....69
10. Never Automate Confusion....79
11. The AI Ready Scorecard™....87

PART THREE — BUILD....98

12. Choosing the Right First Initiative....99
13. Choosing the Right Partner....106
14. Your 90-Day Roadmap....113
15. Prepare People, Not Just Systems....121

PART FOUR — EXECUTE....129

16. Adoption Is Earned, Not Announced....130
17. Are Leaders Actually Leading?....139
18. Stop Using AI Like a Tutor....147
19. Weekly Wins and Removing Barriers....155

PART FIVE — OPTIMIZE....163

20. Transformation Is a Capability, Not a Project....164
21. Scaling and Governing AI....172
22. Ready for What's Next, Not Just What's Now....181

Conclusion: A Letter Before You Go......188
Appendix......190
Notes......201
About the Author......204

PROLOGUE

We Celebrated the Wrong Victory

I remember that meeting as if it were yesterday.

We were sitting around the boardroom table discussing one of the organization's biggest challenges: accounts receivable.

The numbers were difficult to ignore. We weren't invoicing customers accurately enough. Invoices weren't consistently going out on time. Collections continued to grow. An increasing percentage of our customer base had become habitual late payers.

Everyone in the room agreed something had to change.

Then someone spoke up.

"I've been meeting with a company that has an AI tool for collections. I think it could really help us."

Around the room, heads nodded. It sounded innovative. It sounded promising. It sounded like the answer.

But I remember thinking something very different.

Collections wasn't the problem.

Collections was simply where the problem had finally become visible.

The real problem hadn't started weeks earlier. It hadn't even started months earlier.

It had been building for years.

Years of manual processes. Years of disconnected technology. Years of limited visibility into what was really happening. Years of inconsistent data. Years of unclear ownership. Years of workarounds instead of permanent solutions. Years of measuring activity instead of outcomes. Years of leaders making the best decisions they could with incomplete information.

By the time an account reached collections, dozens of decisions had already influenced whether that invoice would ever be paid on time.

Collections wasn't where the problem began.

It was where years of organizational decisions finally became impossible to ignore.

That's why I couldn't understand why we were aiming AI at the last step in the process.

We were asking technology to repair the consequences instead of helping us understand the causes.

A few weeks later, we were back in the same room.

This time there was excitement. The AI initiative had recovered six figures in revenue. People applauded. High-fives were exchanged. The project was declared a success.

Then someone asked, "Where else can we use AI?"

No one asked the question that mattered most.

Did we reduce the percentage of habitual late payers? Did invoice accuracy improve? Were invoices going out on time? Were fewer accounts ever reaching collections in the first place? Had we actually changed the conditions that created the problem?

Those questions were never asked.

Instead, we celebrated the amount collected.

For an organization approaching a billion dollars in annual revenue, recovering six figures without changing the underlying conditions wasn't transformation.

It was a temporary win against a permanent problem.

What frustrated me most wasn't that we chose an AI solution.

It was that no one came back to the people who had identified the real problem in the first place.

I walked into that meeting already knowing where the breakdowns were. The benchmarking was there. The trends were there. The operational evidence was there. The symptoms were simply revealing what had been happening for years.

But when the person responsible for transformation doesn't have the sponsorship to influence decisions…

When they don't have the visibility to connect the dots…

When they don't have the organizational platform to challenge assumptions…

When they don't have the respect, trust, or cultural safety to ask difficult questions...

The organization doesn't just miss opportunities.

It repeats mistakes.

I left that meeting knowing something many people in the room didn't.

We would be back here again.

Because nothing upstream had changed.

The conditions that created the problem were still intact.

That day fundamentally changed how I think about artificial intelligence.

Organizations often say AI didn't work.

But that's rarely the truth.

AI simply reveals what leadership has failed to address.

There wasn't a technology problem. There was a leadership problem. There was a strategic clarity problem. There was a measurement problem. There was a conditions problem.

Organizations often ask AI to solve the place where pain becomes visible.

Transformation leaders learn to ask a different question:

Where was this outcome created?

Every result has an origin. Every symptom has a source. Every organizational problem leaves clues long before it becomes a crisis.

Leaders who learn to find those clues don't just solve problems.

They prevent them.

That realization became the foundation for how I approach transformation, leadership, and artificial intelligence.

And it's why I wrote this book.

My hope is that after reading these pages, you won't just know more about AI.

You'll ask better questions. You'll solve the right problems. And you'll create the conditions where AI can do what it was always meant to do — not replace great leadership, but amplify it.

Because here is the one truth this entire book is built on:

AI will never outperform the conditions of the organization that implements it.

INTRODUCTION

The Technology Was Never the Problem

Every generation of leaders faces a defining moment.

For some, it was the internet. For others, it was cloud computing, mobile, or the wave we simply called "digital transformation."

For you, it's artificial intelligence.

And whether you've embraced it, resisted it, or quietly worried about it, you already know one thing: you don't get to sit this one out.

Organizations around the world are investing billions of dollars in AI. Boards are demanding strategies. Executives are announcing initiatives. Employees are experimenting with new tools — some with permission, many without. The pressure to move is real, and it's growing.

Yet behind closed doors, many leaders share the same quiet frustration:

"Why aren't we seeing the results we expected?"

The usual explanations get offered. The technology isn't mature enough. The employees need more training. We haven't found the right tool yet. Maybe the next vendor will be different.

After years of leading transformation inside organizations — through mergers, turnarounds, system implementations, and cultural change — I've come to a different conclusion.

The technology is rarely the problem.

AI doesn't create organizational dysfunction. It reveals it.

It shines a light on unclear priorities, inconsistent leadership, broken processes, unreliable data, and cultures that were never prepared for change. It exposes the conditions that were already there — the ones that have quietly been shaping your results all along.

Organizations with healthy conditions accelerate.

Organizations with unhealthy conditions struggle faster.

That's why this isn't another book about artificial intelligence.

It's a book about leadership.

Because leaders create the conditions that determine whether AI becomes a catalyst for growth or another expensive disappointment.

Who This Book Is For

If you picked up this book, you're probably carrying responsibility for people, decisions, or an organization you care about.

Maybe you're a CEO who knows AI matters but doesn't have a roadmap.

Maybe you're a board member being asked to approve investments you can't fully evaluate.

Maybe you're a newly appointed executive expected to "figure out AI" for your function.

Maybe you're a transformation leader who has been handed the initiative — and you're worried about adoption, because you've seen what happens when technology arrives before people are ready.

Or maybe you're a business owner watching competitors move and wondering whether you're already behind.

Wherever you're starting from, you have questions:

Where do we begin? Is our organization actually ready? How do we avoid wasting money? How do we get people to embrace this instead of fear it? How do we make it last?

This book answers those questions.

What This Book Won't Do

Let me be honest with you before we go further, because I intend to respect your time on every page.

This book won't teach you how to write code.

It won't rank the latest AI tools — a list that would be outdated before the ink dried.

It won't promise that AI will transform your organization, because it won't. Not on its own.

And it won't sell you hype. You've heard enough of that.

Here's what it will do.

It will give you a clear, repeatable way to lead your organization through AI — before implementation, during implementation, and long after the launch celebrations end. Not just for this initiative, but for every wave of technology that follows. Because there will be more. The tools will change. The questions leaders must ask will not.

Most people are dabbling with AI. This book is about something different: putting AI to work inside your actual business — your goals, your processes, your people — in a way that's sustainable, scalable, and repeatable.

The Journey Ahead

Every organization that adopts AI moves through five phases. Organizations that skip a phase don't move faster. They create risk.

This book walks you through all five.

Part One: Awareness. Why does this matter? We'll start where your people are starting — with the questions, the fears, the hype, and the quiet experimentation already happening in your building. You'll learn why people don't actually resist AI (they resist uncertainty), and why the future belongs to organizations that pair people *and* the machine — never one instead of the other.

Part Two: Diagnose. Are we actually ready? Before you spend another dollar, you'll learn to find the real problem — not just the place where pain becomes visible. You'll examine your leadership, your measurements, and your processes. And you'll complete the AI Ready Scorecard™, an honest assessment of your organization's readiness across the dimensions that determine success.

Part Three: Build. What's the plan? You'll choose the right first initiative, the right partners, and build a 90-day roadmap that prepares your people — not just your systems.

Part Four: Execute. Are we leading it well? Adoption is earned, not announced. You'll learn how to lead visibly, create momentum through weekly wins, remove barriers quickly, and help your people move from using AI like a tutor to working with it like a teammate.

Part Five: Optimize. How do we make it last? This is where most organizations — and most books — stop. You won't. You'll learn how to turn a successful initiative into a lasting capability, so that the next wave of AI finds your organization ready.

Roughly speaking, Awareness and Diagnose are your first thirty days. Build is your second thirty. Execute is your third. Optimize is everything after — because transformation is a capability, not a project.

A Promise Before We Begin

I've sat in the boardrooms where these decisions get made. I've watched organizations celebrate the wrong victories, measure the wrong things, and aim powerful technology at the wrong problems. I've also watched leaders get it right — and I can tell you the difference was never the technology.

So here's my promise to you.

I won't waste your time. I'll challenge your thinking, share what experience has taught me, and give you something you can use — in every chapter, starting Monday morning.

And I'll hold you to one standard throughout this book, the same one I hold myself to:

This is not about perfection. It's about progress.

You will not get everything right. Neither did I. Neither has any organization I've ever worked with. The leaders who succeed with AI aren't the ones who avoid mistakes. They're the ones who create the conditions to learn from them quickly.

Because in the end, everything in this book comes back to a single truth — one I learned the hard way, in a boardroom, watching an organization celebrate a victory that changed nothing:

AI will never outperform the conditions of the organization that implements it.

Let's build those conditions together.

Your ninety days start now.

PART ONE — AWARENESS

"Why does this matter?"

CHAPTER 1
Why AI, Why Now?

The Question Nobody Asked

I've sat in a lot of rooms where big decisions were made. I can tell you about meetings where budgets were fought over line by line, where acquisitions were debated for months, where a single hire was discussed longer than some strategies.

But I cannot tell you about the meeting where we decided why we were adopting AI.

Because it never happened.

There was no debate. No examination. No moment where someone stood at the whiteboard and asked, "What problem are we actually trying to solve, and is AI the right way to solve it?"

What I remember instead is how casually it arrived. AI entered our conversations the way weather enters a conversation — as something happening to everyone, everywhere, that we simply needed to respond to.

The reasoning, when it surfaced at all, came down to two ideas.

Everyone else is doing it.

And: *It could help us reduce staff, which would help our EBITDA.*

That was it. That was the why.

What struck me most was who was in the room. This wasn't a room full of people from one generation, seeing the world through one lens. We had leaders spanning three generations — leaders who had lived through the internet, through cloud, through every wave of "digital transformation" the past thirty years had produced.

And not one of us stopped to have a real "why AI" conversation.

Not one.

AI wasn't examined the way a decision of that size deserved to be examined. It was treated as *the* solution rather than part of one. And because the why was never consistent — cost savings one week, keeping up with competitors the next — the how never had a chance. It was introduced in the least effective way possible, because there was no foundation underneath it.

Looking back, I shouldn't have been surprised by what happened next. When you skip the why, everything downstream inherits the confusion.

What Was Actually Happening

Let me name what was really going on in that room, because I suspect it's going on in yours.

Borrowed urgency was standing in for strategy.

"Everyone else is doing it" feels like a reason. It isn't. It's pressure wearing a costume.

Real strategy starts with your business — your customers, your bottlenecks, your opportunities, your people. Borrowed urgency starts with everyone else's business and works backward, hoping something applies to you. It's the difference between navigating by your own map and following the car in front of you because it looks confident.

Here's the test: if your reason for adopting AI would be equally true for every other company in your industry, it isn't a reason. It's a trend. Reasons are specific. Trends are ambient.

And the only specific "why" in the room was a dangerous one.

I want to be careful here, because efficiency is a legitimate business outcome. There is nothing wrong with wanting a healthier EBITDA. Every executive in that room had a fiduciary responsibility to pursue it.

But when *reduce headcount* is the first, loudest, or only answer to "why AI," something important happens — and it happens whether you announce it or not.

You have just defined AI as a threat to your own people.

Your employees will not read your strategy documents. They will read your intentions. And intentions leak — through what gets funded, what gets measured, what gets celebrated, and what gets whispered after the meeting ends. If the honest answer to "why are we doing this?" is "so we can do this with fewer of you," your people will know. Long before you say a word.

Which means the fear you'll spend the next year fighting, the resistance you'll blame on "change management," the quiet sabotage and slow adoption you'll attribute to culture — you manufactured all of it in that first unexamined moment. The why isn't just the beginning of your strategy.

It's the first message your people receive.

And AI was cast as the solution, not part of one.

Technology has never once fixed a business by itself. Not the internet, not the cloud, not the smartphone, and not AI. Technology is an ingredient. Leadership, process, culture, and strategy are the recipe. When you treat an ingredient as the whole meal, you don't get a meal. You get flour on the counter and a lot of explaining to do.

The Real Why — and the Real Why Now

So if "everyone else is doing it" isn't a why, and "cut overhead" is an incomplete and dangerous why, what does a legitimate one look like?

A real why is anchored to your business, and it usually sounds like one of these:

We have more demand than capacity, and AI can help our existing team serve it.

Our leaders are drowning in low-value work and starving for time to think.

Our customers wait too long for answers we already have.

We make decisions on last month's data in a business that moves daily.

Our best people spend their best hours on work that doesn't require them.

Notice something about every one of those statements. None of them mention a tool. All of them mention your business. And every one of them frames AI the same way: as a way to amplify your people, not a way to eliminate them.

That's not soft-hearted positioning. It's strategy. The organizations getting real results from AI aren't the ones that used it to shrink. They're the ones that used it to build capacity — to grow without breaking, to give their people leverage, to get faster at the things that create value. The savings show up. But they show up as a byproduct of a healthier operation, not as the mission statement.

Then there's the second half of the question: why *now*?

Not because of the hype. The hype is noise.

Now, because AI compounds. The organizations that start learning today — building judgment about where AI helps and where it doesn't, developing people who know how to work with it, making the small mistakes while they're still cheap — will be years ahead of the organizations that wait for the technology to "mature." The tools will keep changing. The organizational learning is what compounds. You cannot buy it later. You can only start it.

And now, because whether you've decided or not, your people already have. We'll get to that in Chapter 5 — but I promise you, AI is already in your building. The only question is whether it's there with your leadership or without it.

The Conversation You Owe Your Organization

Here's what I wish had happened in that room, and what I'm asking you to make happen in yours.

Before a single tool is purchased, before a single vendor demo, before anyone says the word "pilot" — put your leadership team in a room for two hours with no technology on the agenda. Only questions:

What are the three biggest constraints on our growth right now?

Where does our team spend time that doesn't create value for customers?

Where are we slow, and what does slow cost us?

If we had 20 percent more capacity tomorrow, where would we point it?

What would we never want AI to touch — and why?

If our employees read our real reasons for adopting AI, would they lean in or lock down?

You'll notice AI barely appears in those questions. That's the point. The why-AI conversation is really a know-your-business conversation. AI just happens to be the occasion for it.

If your team can answer those questions with clarity and consistency, you have the beginnings of a real why — one you can say out loud, in front of your entire workforce, without flinching. That's the test. A why you can't say to your own people isn't a why. It's a secret. And secrets don't scale.

If your team can't answer them yet, congratulations — you just saved yourself an enormous amount of money. Because you found out you weren't ready *before* the invoices arrived, not after.

Either way, you're ahead of the room I sat in.

The Executive Brief

Key Insight

"Everyone else is doing it" is borrowed urgency, not a strategy. Your why must be anchored to your business — and it's the first message your people receive, whether you announce it or not.

Questions Every Leader Should Ask

- Can every member of my executive team state, consistently, why we're adopting AI?
- Would I be comfortable saying our real reasons out loud to our entire workforce?
- Is our why anchored to a specific business problem, or to what competitors are doing?

- If headcount reduction is part of our thinking, have we considered what that why will cost us in fear, resistance, and adoption?
- Are we treating AI as the solution, or as part of one?

Monday Morning Actions

1. Schedule the two-hour "no technology on the agenda" conversation with your leadership team.

2. Ask each executive to independently write one sentence: "We are adopting AI because ______." Compare the answers. The gaps are your starting point.

3. Identify the three biggest constraints on your growth. Hold them — you'll use them in Part Two.

Your AI Prompt

"Act as a skeptical board member. I'm going to tell you why my organization is adopting AI. Challenge my reasoning: identify where my 'why' is borrowed from industry trends rather than anchored in my specific business, where it's vague, and where it might create fear in my workforce. Then help me sharpen it into one clear statement I could confidently share with every employee. Here's my current thinking: [describe your reasons]."

Reflection

Think about the last significant technology decision your organization made. Was there a real "why" conversation — or did the decision arrive the way weather arrives? What did that cost you?

Your 90-Day Checkpoint

By the end of this week, you should be able to write your organization's why for AI in one sentence — a sentence anchored in your business, consistent across your leadership team, and safe to say in front of your people. If you can't yet, stay in this chapter. The rest of the book will still be here.

CHAPTER 2
Why AI Isn't Delivering What Leaders Expected

Two Years of Talking

When leaders tell me AI isn't delivering, they usually mean one of two things.

The first is the story you'd expect: they bought the tools, launched the initiative, made the announcement — and the results underwhelmed. We'll spend plenty of time on that failure in this book.

But there's a second story, and it's far more common than anyone admits. In this one, AI didn't deliver because it never got the chance.

Let me tell you about a problem I watched an organization live with.

We had significant challenges reconciling payments. If you've ever been near an accounts receivable function, you know exactly what that means — and how far the damage travels. It isn't just an accounting inconvenience. It's cash flow you can't see clearly. It's customers whose on-time payments get recorded late, damaging relationships and trust through no fault of their own. It's decisions made on numbers nobody fully believes.

The transformation team understood the problem. We also understood something else: this was almost a perfect use case for AI. Reconciliation is high-volume pattern matching — connecting payments to invoices across thousands of transactions, learning from exceptions, getting better over time. It's tedious, repetitive, and error-prone for humans, and tailor-made for a technology that never gets tired and never stops learning.

We weren't guessing about the solutions, either. We were actively vetting tools. Every serious vendor in the space — including the ones we'd identified as the best potential partners — had built AI into their platforms. The capability existed. It was sitting in demo environments, waiting.

And then leadership would ask the question.

"What if the people can't do it? How can we do it with this tool?"

I heard versions of that question again and again. If our own team can't get reconciliation right, how could a piece of software possibly do it?

To this day, I'm not sure whether the gap was a lack of education or a lack of understanding. Maybe there's no difference. But I know what the gap cost.

Two years later, we were still talking about the same problem. Still vetting. Still asking the same questions. Still reconciling by hand.

We never got there.

What Was Actually Happening

I want to sit with that leadership question for a moment, because it sounds reasonable — and that's exactly what makes it dangerous.

"If our people can't do it, how can a tool?"

Underneath that question is an assumption almost every leadership team carries into their first AI conversation: that AI is essentially another employee. That it can only do what people can do, maybe a bit faster. That if a task defeats your team, it will defeat the software too.

That assumption is backwards.

AI is not a faster version of your people. In work like reconciliation, it is strong precisely where humans are weak. A person matching payments to invoices gets tired at transaction two hundred. Loses focus after lunch. Misses the pattern that only appears when you can hold ten thousand transactions in view at once. AI doesn't. The very reasons your people struggled with the work were the reasons the technology was suited to it.

The leaders in that room weren't asking a bad-faith question. They were asking the only question their understanding allowed. And that's the real lesson:

You cannot evaluate what you don't understand. So you delay it instead.

Nobody in that room ever said "no" to solving reconciliation. Nobody killed the initiative. There was no meeting where we decided not to act. There was just... more vetting. More questions. More conversations that felt like diligence and functioned like avoidance.

That's what makes this failure mode so hard to see. A failed implementation shows up in a budget review. Somebody has to answer for it. But paralysis? Paralysis looks responsible. It looks careful. It photographs well in a boardroom.

And it costs more.

The Invoice That Never Arrives

Here's the accounting nobody did during those two years.

Every month we didn't fix reconciliation, cash sat unrecognized or misapplied. Every month, customers who paid on time were recorded as if they hadn't — and some of them noticed. Every month, the team spent hours on manual matching that a system could have learned to do, hours that could have gone to the exceptions and judgment calls that actually needed human attention. Every month, decisions were made on receivables data that everyone quietly discounted.

None of that ever arrived as an invoice. No one had to sign for it. There was no line item called "the cost of still talking about this."

But it was real, and it compounded — at the same rate the unreconciled accounts did.

That's the pattern I want you to see, because it's not about reconciliation. Failed AI initiatives get autopsies. Deferred ones get amnesia. Organizations tally every dollar spent on the tool that disappointed them and zero of the dollars lost to the tool they never deployed. Which means when leaders say "AI isn't delivering," they're only counting one of the two ways it fails.

Inaction doesn't send an invoice. That doesn't make it free.

One Gap, Two Failures

If you read Chapter 1, you might notice something.

In that room, leadership moved on AI without a real why — borrowed urgency, no examination, adopted in the least effective way.

In this room, leadership had a genuine why — a clearly defined, high-value, well-suited problem — and couldn't move at all.

Rushing and freezing look like opposite failures. They're the same failure. Both rooms were full of capable, experienced leaders making decisions about something they didn't sufficiently understand. One room's answer was to charge ahead. The other's was to keep asking questions that couldn't be answered because the questioners didn't know what a good answer would look like.

The gap in both cases wasn't technology. The vendors had the technology. The gap wasn't the people. The teams were capable. The gap was executive understanding — and here is the uncomfortable truth underneath this entire book:

You cannot lead what you do not understand.

Notice what I'm not saying. I'm not saying you need to become technical. You don't need to write code, understand model architectures, or keep up with every release. Executives lead finance without doing the bookkeeping and lead legal without passing the bar. But they understand what those functions are good at, what they're bad at, and what questions to ask them.

AI deserves the same literacy. At minimum, that means understanding three things:

What AI is genuinely good at — pattern recognition at scale, drafting, summarizing, matching, monitoring, learning from volume. Work that is high-repetition, high-volume, and rule-adjacent.

What AI is genuinely bad at — judgment without context, accountability, relationships, knowing what it doesn't know, and understanding your business unless you teach it.

What questions actually evaluate it — not "can it do what our people do?" but "what does this work actually require, and which of those requirements are pattern and volume, and which are judgment and trust?"

If the leaders in my story had held just that much literacy, the question changes. "What if the people can't do it?" becomes "The people can't do it *because it's not human-shaped work* — which is exactly why we should test the tool." Two years becomes ninety days.

That literacy is what the rest of this book builds. But it starts with an honest admission that most executive teams never make: *we are being asked to lead something we don't yet understand — and our first investment should be in understanding it.*

The room I sat in never made that admission. So the questions never got better. And the problem outlived two years of meetings about it.

Don't let that be your organization. The most expensive AI decision you can make isn't the wrong tool.

It's the third year of talking.

The Executive Brief

Key Insight

AI fails to deliver in two ways: implementations that disappoint, and decisions that never get made. Both come from the same gap — leaders evaluating something they don't yet understand. Failed initiatives get autopsies; deferred ones get amnesia. Only one of them appears in your budget, but both compound.

Questions Every Leader Should Ask

- Is there a well-defined, high-value problem we've been "vetting" solutions for over multiple quarters? What has that delay cost us — in cash, customers, and hours?

- When we evaluate AI, are we asking whether it can do what our people do — or what the work itself actually requires?
- Does our leadership team have enough AI literacy to know what a good answer to our questions would even look like?
- Where in our organization does caution get credit while its costs go uncounted?

Monday Morning Actions

1. List every technology decision your organization has been "evaluating" for more than six months. For each, estimate the monthly cost of not deciding — then decide, or set a decision date.

2. Pick your longest-standing operational headache. Ask: is this work primarily pattern-and-volume, or primarily judgment-and-trust? If it's the first, it belongs on your AI shortlist.

3. Schedule two hours of AI literacy for your executive team — not a vendor demo, an education session. What AI is good at, bad at, and how to evaluate it.

Your AI Prompt

"I'm going to describe a business process my organization struggles with. Break it down into its component tasks, and for each task, tell me whether it is primarily high-volume pattern work (well-suited to AI), primarily judgment and relationship work (should stay with people), or a mix. Then tell me what a people-AND-machine version of this process could look like. Here's the process: [describe it]."

Reflection

Think of a decision your organization delayed for a year or more. When it was finally made — or quietly abandoned — did anyone ever calculate what the waiting cost? Why not?

Your 90-Day Checkpoint

By the end of this week, identify one problem in your organization that has been discussed longer than it would take to test a solution. Write down what the delay has cost. You'll carry this into Part Two, where you'll diagnose whether it's the right place to start.

CHAPTER 3
The Fear Nobody Names

"No One Is Ready"

A few years ago, I attended a seminar that changed how I think about AI adoption — and it wasn't a seminar about technology.

The presenters shared something that stopped me: IT professionals were struggling. Not with workloads or systems. With fear. More of them were seeing therapists than ever before. Their sense of psychological safety was under attack. And at the center of it was a single belief spreading through the profession: *AI is coming for my role.*

Think about who we're talking about. These weren't people on the margins of technology, afraid of what they didn't understand. These were the people closest to it — the ones who would be asked to evaluate the tools, implement the systems, and carry every AI initiative their organizations dreamed up. And they were quietly coming apart.

I remember sitting in that room thinking: *I can get ahead of this.*

So I went to the leader I reported to at the time with a simple idea. Let's hold an open meeting with the IT team. Not a meeting about AI — about them. How are they feeling? What are they already using? What do they know? What don't they know? Get it all on the table and start having honest conversations.

My reasoning was straightforward. If the people we're counting on to lead the rollout are living in fear, we don't have a technology problem coming. We have a trust problem coming — and it will touch everything.

The idea was discounted in a sentence.

"No one is ready for that."

And the conversation moved on.

I've turned that response over many times since. Was it that other priorities had taken precedence? That my concern didn't seem valuable? Or that this leader simply didn't believe the people they served were struggling? I'll never know. What I know is that the meeting never happened. The fear never got named.

And here's what has stayed with me: *"No one is ready"* was said as a reason not to have a conversation about readiness. Read that again. The fear in that organization wasn't only in the IT team.

It was in the room where I made the suggestion.

What Was Actually Happening

There were two fears operating that day, and only one of them belonged to the employees.

The first fear — the one the seminar named — is the one you'd expect. People were afraid of being replaced. Afraid of becoming irrelevant. Afraid that the skills they'd spent careers building were about to be devalued by something they didn't choose and couldn't control.

I want to say something clearly about that fear, because too many leaders treat it as a problem to be managed: it is a *reasonable response to genuine uncertainty.* Your people are not being dramatic. They read the same headlines you do. They've watched organizations — maybe yours — talk about AI in the same breath as "efficiency" and "overhead." If your why for AI was never examined, never articulated, or never safe to say out loud (Chapter 1), your people filled the silence with the worst version of the story. That's not irrationality. That's what humans do with uncertainty.

And if you're tempted to file what I heard at that seminar under "anecdote," the data won't let you.

The American Psychological Association's Work in America research found that nearly four in ten U.S. workers worry AI may make some or all of their job duties obsolete. Now look at what that worry does to people. Among workers worried about AI, roughly half said their work negatively impacts their mental health — compared to fewer than a third of workers without that worry. Nearly two-thirds of AI-worried workers reported feeling tense or stressed during a typical workday, versus just over a third of everyone else.

Sit with that comparison for a second. Same organizations. Same jobs. Often the same workloads. The difference wasn't the work.

It was the uncertainty.

And it isn't fading as AI becomes familiar — it's growing. Pew Research Center found over half of U.S. workers worried about AI's future impact in the workplace, with the worried outnumbering the hopeful. Mercer's global survey of twelve thousand workers found the share who fear losing their job to AI climbing sharply in just two years. Familiarity isn't curing this. Silence is feeding it.

Two more findings deserve a seat at your next executive meeting.

First: APA's research found middle managers and frontline workers among the *most* likely to fear their duties becoming obsolete. Read that carefully. The people most afraid are the exact layers of your organization you will depend on to carry adoption when we get to Part Four. Fear isn't concentrated at the edges of your org chart. It's concentrated in the middle of your execution engine.

And there's a reason the middle carries the most fear: it also carries the most load. Research out of Harvard Business Review found that AI adoption is quietly overloading middle managers — the layer expected to implement the technology, maintain quality, and develop their teams through the change, all at once, usually on top of an untouched workload. Fear and load converge on the same people — so whatever you ask of your middle layer will need to be funded with time, support, and updated expectations, not just announced.

Second: workers worried about AI were significantly more likely to say they intended to look for a new job within the year. Fear doesn't just make people anxious. It makes them leave — and the ones with the most options leave first.

Here's the finding I want you to hold onto, though, because it validates everything this chapter is asking of you: the APA's own guidance for employers is that transparent, honest communication about AI — educating people, involving them, giving them a voice in the process — measurably reduces the stress. In other words, the prescription for AI anxiety was never a wellness program or an extra day off.

It's leadership. It's the conversation.

The exact conversation, as it happens, that I proposed — and that got answered with "no one is ready."

Which brings us to the principle at the heart of this chapter:

People don't resist AI. They resist uncertainty.

I've watched employees embrace enormous change when they understood where they stood in it. And I've watched modest, sensible changes collapse because nobody would tell people the truth about what was coming. The variable was never the size of the change. It was the size of the silence around it.

The second fear — the one nobody names — belonged to leadership. When my leader said "no one is ready," I don't believe that sentence was really about the IT team. Leaders avoid these conversations for their own reasons: they're afraid of opening a door they can't close, of being asked questions they can't answer, of making promises they can't keep — or of discovering that the fear is worse than they hoped. Avoiding the conversation feels like containing the problem.

It isn't. It's outsourcing the conversation to the rumor mill.

Because here's the thing about unnamed fear: it doesn't wait for permission, and it doesn't stay put. It converts. Into silence in meetings where you needed honesty. Into your best people quietly updating their résumés. Into polite non-adoption, where everyone attends the training and nothing changes. And into quiet, unsanctioned experimentation, as people privately race to make themselves indispensable — which is one of the roads to Shadow AI, and exactly where Chapter 5 will pick up.

The fear conversation happens either way. The only decision a leader gets to make is whether to be in the room for it.

The Meeting That Should Have Happened

I never got to run the meeting I proposed. So let me give it to you instead — because it's the meeting your organization probably needs, and it costs you ninety minutes.

The setup matters more than the agenda. This is not an AI briefing. No slides about tools. No roadmap reveal. The stated purpose is the honest one: *we're going to be living with this technology for the rest of our careers, and before we make plans, I want to understand where everyone actually is.* If it's a large organization, start with the team closest to the rollout — because their fear travels furthest.

Four questions, asked in this order:

How are you actually feeling about AI — honestly? Then wait. The first silence will be long. Do not fill it.

What are you already using? Ask this without judgment attached, because you want the truth and the truth may include tools nobody approved.

What do you know that we should know? Your people are closer to the work than you are. Some of them are further ahead on AI than the executive team. Let that surface.

What don't you know that worries you? This is where the real fears live — and where you learn exactly what your communication has failed to address.

And then the hard part: what you say back. This is where most leaders fail, because they reach for reassurance they haven't earned. Hear me on this: *do not make promises you cannot keep.* "No one will lose their job" is a sentence that feels compassionate in the moment and detonates later if it turns out not to be true — taking your credibility and every future change initiative with it.

You can be honest and steadying at the same time. It sounds like this: *Here's what I know. Here's what I don't know yet. Here's what we're committed to — that nobody will be surprised, that we'll invest in your ability to work with this technology, and that this conversation isn't a one-time event.* Certainty is not the standard. Candor is. People can stand almost any truth. What they can't stand is discovering they were managed instead of told.

One more thing. If you're reading this as the leader who would be *receiving* that meeting proposal — the one with the power to say "no one is ready" — I'd ask you to sit with a question before you say it: *Is it them who isn't ready? Or is it me?* There's no shame in the honest answer. There's only cost in the dishonest one.

The Cost of the Meeting That Doesn't Happen

I don't know exactly what the silence cost that organization, because the costs of silence never report to anyone. But I know what it didn't buy.

It didn't buy readiness. The fear the seminar described didn't dissolve because we declined to discuss it. It didn't buy time. The uncertainty compounded, the way it always does. And it didn't buy trust — because people always know when a conversation is being avoided, and they draw their own conclusions about why.

Fear, it turns out, is not the enemy of AI adoption. Fear is *data*. It tells you precisely where your communication has gaps, where your why hasn't landed, and where trust needs rebuilding before technology arrives. Every fear your people name is a problem you get to solve before it becomes a resistance you have to fight.

The enemy was never the fear.

The enemy is the silence around it.

The Executive Brief

Key Insight

People don't resist AI — they resist uncertainty. Unnamed fear doesn't disappear; it converts into silence, attrition, non-adoption, and shadow behavior. The fear conversation happens with or without you. Leadership is deciding to be in the room.

Questions Every Leader Should Ask

- When did we last ask our people how they feel about AI — not what they think of a tool, but how they *feel*?
- What story are our employees telling themselves about why we're adopting AI? Do we know, or are we assuming?
- Is the team closest to our rollout confident or afraid? How do we actually know?
- What conversation am I avoiding because *I'm* not ready — and what is that avoidance teaching my organization?
- Have we made, or implied, any promises about AI and jobs that we may not be able to keep?

Monday Morning Actions

1. Schedule the ninety-minute meeting described in this chapter with the team closest to your AI plans. Four questions. No slides.

2. Before the meeting, write down what you believe your people fear most. Afterward, compare it to what they actually said. The gap is your communication debt.

3. Draft your honest answer to "will AI cost people here their jobs?" If you can't answer it candidly yet, that's your first executive conversation — because your people are already answering it for you.

Your AI Prompt

"Act as an experienced organizational psychologist. I'm preparing to hold an open conversation with my team about their feelings, fears, and experiences with AI. Help me prepare: give me follow-up questions for when answers are

guarded, guidance on responding to fears I can't resolve, and phrases to avoid because they sound like managed reassurance rather than honesty. My team's context: [describe your team]."

Reflection

Think about a time in your career when a leader avoided a conversation everyone knew needed to happen. What did the silence teach you about what — and who — mattered? Now ask: what is your silence on AI teaching your people?

Your 90-Day Checkpoint

By the end of next week, you should have held at least one honest fear conversation and be able to name your organization's top three AI fears in your people's own words — not your assumptions. You'll need them: the next chapter is about the answer to those fears, and Part Four's adoption work is built on what you learn here.

CHAPTER 4
The People AND the Machine

The Sentence That Changed the Room

Not every lesson comes from a boardroom. Sometimes it comes from a webinar you almost didn't attend.

I was watching a session on AI in the workplace — one of dozens I sat through during those years, most of them forgettable. And then the speaker referenced a line that made me stop taking notes.

"AI won't take your job. It's somebody using AI that will take your job."

The line came from economist Richard Baldwin, speaking at the World Economic Forum's Growth Summit in 2023, and it has traveled the world since — for good reason.

I remember thinking: *that's golden.* Not because it was clever. Because in one sentence, it did what months of corporate reassurance could never do.

I had just spent years watching organizations fumble the AI conversation. I'd sat in the seminar where they described IT professionals in therapy over replacement fears. I'd watched a leader decline the honest conversation because "no one was ready." Everywhere I looked, the message people were absorbing was the same: *the machine is coming, and there's nothing you can do about it.*

And here was a sentence that quietly dismantled that message.

Because look at what it actually does. It takes the threat out of the machine's hands and puts the choice in yours. You cannot control what AI becomes. You cannot vote on what your industry adopts. But whether you learn to work with it? That's yours. Entirely.

The fear says: *something is happening to me.*

The sentence says: *something is available to me.*

That shift — from victim of the technology to user of it — is the entire difference between a workforce that resists AI and a workforce that runs toward it.

Handle With Care

Before we go further, I need to say something about that sentence, because I've since watched leaders get it badly wrong.

Delivered one way, it's an invitation. Delivered another way, it's just fear with a new address.

Imagine an executive standing in front of an anxious workforce and saying, in effect: *"Don't worry — the machine won't replace you. Your coworker will."* All that leader has done is redirect the threat from the technology to the person at the next desk. Same fear. New target. And now you've added internal competition to the anxiety you were trying to cure.

The sentence only works as an invitation — and invitations require something from the host.

If you tell your people the future belongs to those who learn AI, then you have taken on three obligations in the same breath. You owe them *access* — real tools they're permitted to use, not a policy that forbids the very learning you just demanded. You owe them *time* — because "learn AI on top of everything else" is not an invitation, it's a burden with a bow on it. And you owe them *safety* — permission to be clumsy at first, to ask basic questions, to experiment without their competence being judged, which is exactly the psychological safety we talked about in Chapter 3.

Say the sentence and pay for it, and you've handed your people back their agency.

Say the sentence and pay for nothing, and you've simply found a more sophisticated way to frighten them.

The False Choice

Here's why that webinar moment belongs in this book, though — because underneath the sentence sits the reframe this entire chapter exists to make.

Almost every failed AI conversation I've witnessed rests on the same unexamined assumption: that this is a contest. People *or* the machine. Somebody wins, somebody goes home.

You can hear that assumption on both ends of the spectrum.

On one end sits the automation dream — the leaders we met in Chapter 1, for whom AI's chief attraction was doing the same work with fewer people. People *or* machine, and they've chosen the machine. We've already seen what that why costs: fear, resistance, and an adoption problem manufactured before the first tool arrives.

On the other end sits refusal — the teams and leaders who hear "people or machine" and choose the people, full stop. No tools, no experiments, no conversation. It feels loyal. It feels protective. And it quietly guarantees that their people fall behind the ones who learned — which, if you take Baldwin's sentence seriously, is the one outcome loyalty was supposed to prevent.

Two opposite postures. One shared mistake. Both accepted the premise that this is a competition.

It was never a competition. It's a pairing.

The future doesn't belong to the people or the machine. It belongs to the people AND the machine — deliberately paired by leaders who understand what each one is for.

What Each One Is For

I want to make the AND concrete, because "humans and AI working together" has become the kind of phrase people nod at and nobody acts on.

Go back to the reconciliation problem from Chapter 2. Thousands of payments, thousands of invoices, patterns hiding in volume no human could hold in view. The machine's work — precisely because it is high-volume, repetitive, and pattern-shaped. The very qualities that exhausted our people were the qualities that suited the technology.

But now look at what's left when the machine takes the matching: the exceptions. The customer whose payment doesn't fit any pattern because their situation doesn't fit any pattern. The judgment call about a long-standing account. The phone call that protects a twenty-year relationship. The decision about what the anomaly *means* and whether policy should change because of it.

That's the human work — and notice something about it. It isn't the leftovers. It's the part that was always the point. Your people didn't build careers to match line items; they built judgment, context, and relationships that no system can replicate. The machine handling the volume doesn't shrink their work.

It returns their work to them.

That's the pattern, and it repeats far beyond finance. The machine brings volume, speed, stamina, and pattern recognition at a scale no human can match. Your people bring judgment in ambiguous situations, accountability for decisions, knowledge of your business and your customers, relationships, and the wisdom to know when the pattern is wrong. Neither one is complete. Ask the machine for judgment and it will give you confident nonsense. Ask your people for ten thousand perfect comparisons before lunch and you'll get burnout and errors.

The question leaders keep asking — *"Can the machine do this job?"* — is the wrong question. Jobs aren't the unit. Tasks are. The right question is: *What does this work actually need — volume or judgment, pattern or relationship — and have we paired the person and the machine accordingly?*

Replacement isn't a destiny that arrives with the technology. Replacement is a design choice. So is amplification. The same tool, in the same organization, can shrink your people or enlarge them — and the difference is not in the software. It's in the leader who decided what the pairing was for.

Which is why I'll say now what this book will keep proving in the chapters ahead: every adoption problem you will face in Part Four, every sustainability question in Part Five, comes back to whether your people believe the AND. Not whether you announced it.

Whether they believe it.

They'll believe it when the machine takes the work that was grinding them down and leaves them the work that made them proud. They'll believe it when the invitation to learn comes with the time and safety to accept it. They'll believe it when the first person who masters the tools gets celebrated and grows — instead of quietly becoming the reason two colleagues were let go.

People don't adopt a technology.

They adopt what the technology proves about their leaders' intentions.

The Executive Brief

Key Insight

The AI conversation fails when it's framed as a contest — people or the machine. The future belongs to people AND the machine, deliberately paired: volume, speed, and pattern to the technology; judgment, relationships, and accountability to your people. Replacement is not a destiny. It's a design choice.

Questions Every Leader Should Ask

- In our AI plans, are we asking "can the machine do this job?" — or "what does this work actually need, task by task?"
- If the machine took the repetitive work out of our teams tomorrow, do we know what higher-value work their time would return to? Have we told them?

- Have we invited people to learn AI — and if so, have we paid for the invitation with access, time, and safety?
- Is there anywhere in our messaging where "learn or be left behind" is functioning as a threat rather than an invitation?
- What has our organization's behavior — not our communications — proven so far about whether we believe in the AND?

Monday Morning Actions

1. Pick one role in your organization. List its ten most common tasks and sort them: pattern-and-volume work (machine-suited) versus judgment-and-relationship work (human work). Share the exercise with the person in that role and compare answers.

2. Audit your invitation. If you've told people to build AI skills, verify all three payments: tools they can access, time carved out to learn, and explicit safety to be a beginner. Fund whichever is missing.

3. Find one early success story where AI made someone's work bigger — more valuable, more interesting — and tell that story publicly. Your people are watching what happens to the first adopters. Make sure they see growth, not risk.

Your AI Prompt

"Here is a list of tasks that make up a role on my team: [list them]. For each task, classify it as (a) high-volume pattern work suited to AI, (b) judgment, relationship, or accountability work that should remain human, or (c) a hybrid where AI drafts and a person decides. Then describe what this role could look like redesigned around that pairing — and what new, higher-value work the freed-up time could support."

Reflection

Think about the most tedious season of your own career — the work that drained you without using what you were best at. If a technology had taken that work off your plate, would you have experienced it as a threat or a gift? What would have made the difference? Now ask: which one are you currently offering your people?

Your 90-Day Checkpoint

By the end of this week, you should be able to articulate the AND for your organization in one sentence: what the machine will carry, what your people will keep, and what their freed capacity will build. If your answer to that last part is "cost savings," go back to Chapter 1 — your why isn't finished yet.

CHAPTER 5

Shadow AI Is Already in Your Building

A Day a Week

I want to tell you about a conversation with a colleague — someone working at a large, well-known energy company.

We were talking shop, the way colleagues do, and the subject turned to AI. Their company had done what most companies do: leadership had selected an AI tool, commissioned it, rolled it out. Official. Sanctioned. Done.

They weren't using it.

They were using something else — a tool nobody had approved, nobody had vetted, and nobody in leadership knew existed inside the company's walls.

Now, here's where the story usually goes in the corporate imagination: the rogue employee, the policy violator, the security risk who needs to be found and corrected. And here's what was actually true.

My colleague wasn't being rebellious. There was no defiance in it at all. They were using the other tool because *it worked better.* It was stronger. It let them do parts of their job they simply couldn't do otherwise. And it was saving them approximately a day a week — a day they were pouring straight back into being more productive at that same company.

Think about what that means. One of this organization's capable, motivated employees had found a way to give their employer twenty percent more of themselves. And the organization had no idea.

How did it come to this? The way it usually does. Somewhere upstream, someone had decided what everyone needed — one person, or one committee, choosing one tool for the entire organization — without ever finding out what the work actually required. The sanctioned tool was an answer to a question nobody had asked the people doing the jobs.

So the people doing the jobs answered it themselves.

And because the organization wasn't taking care of this — wasn't asking, wasn't listening, wasn't providing — its proprietary information was flowing, day after day, into a tool it had no governance over and no visibility into. Company data. Company knowledge. Company intellectual property. Sitting in a system nobody in leadership had ever heard of.

This is not a rare story. I'd argue it's the most common AI story in business today. It is happening every single day, in almost every organization — very likely including yours, right now, while you read this.

The only question is whether you know about it.

What Was Actually Happening

The instinct, when leaders discover Shadow AI, is to treat it as a compliance failure. Tighten the policy. Block the sites. Send the memo. Find the violators.

I want to talk you out of that instinct — not because the risk isn't real (we'll get to the risk, and it's serious), but because the compliance framing misreads what you're looking at.

Shadow AI is not a discipline problem. It's a message.

Go back to my colleague. What does that behavior actually tell the organization, if it's willing to read it?

It tells them their sanctioned tool isn't good enough — that the gap between what they provided and what the work requires is at least a day a week wide. It tells them their selection process was broken: a decision made *for* the workforce instead of *with* it. It tells them exactly where AI creates real value in their business — my colleague found a genuine use case, proved it, and measured it, for free. And it tells them they employ exactly the kind of person every AI initiative dreams of: curious, self-directed, already fluent, already delivering results.

Shadow AI is a requirements document, written by your most motivated employees, describing precisely what your official tools failed to give them. Most organizations burn that document and punish its authors.

And notice who writes it. Not your laggards. Not your resisters. Shadow AI comes from the *front* of your adoption curve — the very people Chapter 4 said would define your future, the ones learning to work with the machine. Remember the bell curve from Chapter 3: on one tail, people frozen in fear and resistance. On this tail, people racing ahead without you. They look like opposite problems.

They are the same problem. Both tails are what a workforce does when leadership is absent from the middle — when nobody has named the fear, provided the tools, or built the channel. Resistance is what the anxious do with your silence. Shadow AI is what the capable do with it.

The Part That Should Keep You Up at Night

Now the risk — because I don't want the sympathy in this chapter to be mistaken for permission.

Every day my colleague used that tool, the company's information went along. Not because my colleague was careless, and certainly not malicious — but because that's how these tools work. What you type in, you have shipped out. Customer details. Contract language. Operational data. Strategy. Pricing. The accumulated, unwritten knowledge of how the business actually runs.

Into a system the company had never evaluated. Under terms nobody in legal had read. With retention policies nobody could name, security nobody had verified, and exposure nobody could measure.

Here's what makes Shadow AI different from most risks on your register: *you cannot see it.* A failed project shows up in a budget review. A breach shows up in an incident report.

Shadow AI shows up nowhere — no line item, no dashboard, no log — until the day it shows up everywhere.

And this is where the standard response makes everything worse. The crackdown — block, forbid, threaten — doesn't end Shadow AI. It has never ended Shadow anything; IT leaders fought this exact war over unsanctioned software for two decades and lost. What the crackdown actually accomplishes is driving the behavior deeper underground. The tools move to personal devices and personal accounts. The time savings continue, because people do not voluntarily hand back a day of their week. But now the one thing you had a chance of getting — *visibility* — is gone for good.

You cannot govern what you cannot see. And you cannot see what you've made dangerous to admit.

Which means the goal is not zero Shadow AI. The goal is zero *reason* for it.

Bringing It Into the Light

So what does a leader actually do? Four moves, in order — and the order matters.

First: find out what's true, and make the truth safe. This is the "what are you already using?" question from Chapter 3's meeting, and it only works under one condition: amnesty. Declare it plainly — for the next sixty days, nobody gets punished for honesty about the tools they're using. You are not asking so you can catch people. You are asking because you need the map. If people fear the answer will be used against them, you will get a clean survey and a dirty reality. In my experience, leaders are consistently stunned by what surfaces — both how much is in use, and how good some of the use cases are.

Second: read the requirements document. Every shadow tool in your building is telling you something specific: what it does that your sanctioned tools don't, who needed it, and what it's worth — my colleague's was worth a day a week, per person, measurable. Sit with your early adopters and let them show you. You will learn more about where AI creates value in your business in those conversations than in any vendor demo you'll ever attend. The people who went around your process are, inconveniently, your best consultants on what the process should have produced.

Third: close the gap they revealed. If the unsanctioned tool is genuinely better, that's not awkward — that's actionable. Evaluate it properly. Buy it, or something that matches it, under enterprise terms with real data protections. If it can't be secured, then meet the *need* it served with something that can be. The one move you cannot make is forcing people back to a tool they've already proven inferior. They won't go. They'll just stop telling you.

Fourth: build the fast lane. Shadow AI grows in the gap between how fast your people move and how fast your organization decides. If requesting a tool takes six months and installing one takes six minutes, your governance has already lost — you've made the unsanctioned path the only rational one. So build a standing, rapid process: any employee can propose a tool, evaluation happens in weeks not quarters, and the answer comes back quickly — approved, approved with guardrails, or no with a reason and an alternative. Governance that moves at the speed of the work gets used. Governance that doesn't gets routed around.

Do these four things and something remarkable happens: your Shadow AI problem converts into your adoption program. The hidden use cases become your pilot portfolio. The rule-breakers become your champions — the exact people Part Four will need. And the day a week my colleague was saving stops being a secret and starts being a strategy.

That company never got any of that. As far as I know, it still doesn't know what it doesn't know — not what my colleague was using, not what that tool could do, not what left the building. It thinks its AI initiative is the tool it commissioned.

Their real AI initiative started without them.

Don't let yours.

The Executive Brief

Key Insight

Shadow AI isn't rebellion — it's a requirements document written by your most motivated employees, describing what your official tools failed to give them. The risk is real (your data in systems you can't see), but crackdowns don't end it; they just end your visibility. The goal is not zero Shadow AI. It's zero reason for it.

Questions Every Leader Should Ask

- What AI tools are actually in use in our organization right now — and how confident am I in that answer?
- Is it safe for an employee to admit they're using an unsanctioned tool? What would happen to them if they did?
- How was our sanctioned AI toolset chosen — with the people who do the work, or for them?
- How long does it take an employee to get a new tool evaluated and approved here? Is the unsanctioned path faster? By how much?
- What proprietary information could already be sitting in tools we've never evaluated?

Monday Morning Actions

1. Declare a sixty-day amnesty and run the discovery: what's in use, for what, and what it's worth. Make the no-punishment commitment explicit and keep it.

2. Ask your leadership team to write down the three AI tools they believe are most used in your organization. Compare against what discovery finds. The gap is your visibility problem, measured.

3. Time your own tool-approval process from request to decision. If it's measured in months, charter the fast lane this quarter — weeks, not quarters, with every "no" accompanied by a reason and an alternative.

Your AI Prompt

"Act as an enterprise AI governance advisor. Help me design a lightweight, fast-moving process for evaluating employee-requested AI tools. Requirements: employees can propose tools easily, evaluation covers data security, legal terms, and business value, decisions come back within [X] weeks, and every rejection includes a reason and a sanctioned alternative. My organization's context: [size, industry, current approval process]."

Reflection

Somewhere in your organization is an employee saving a day a week with a tool you've never heard of. Before anything else, ask yourself honestly: is your first instinct to stop them — or to learn from them? What does that instinct say about the conditions you've created?

Your 90-Day Checkpoint

By the end of this month, you should have a real map of AI use in your organization — sanctioned and shadow — and at least one shadow use case promoted into a properly governed pilot. If your map still matches your policy exactly, you don't have a clean organization. You have a quiet one. Keep asking.

CHAPTER 6
Communicate Why Before What

The Morning the Words Disappeared

I was brought into an organization to help align its operational systems — and its language.

The company had grown through acquisition. Multiple organizations purchased over time, each arriving with its own systems, its own processes, and — this is the part people underestimate — its own vocabulary. My charge wasn't simply to get everyone onto one platform. It was to get everyone into one *language*. Because here's a truth every integration veteran learns the hard way: different words mean different things to different people. The same term can describe two different processes in two acquired companies; two different terms can describe the same one. And if you don't reconcile the language, you don't just confuse people — you corrupt your data, because clean data depends on shared meaning. It was a huge undertaking.

But the real education came from what I found when I arrived.

This organization, I discovered, struggled — badly — with telling its people about change. Before I got there, system cutovers had been happening in a way I still find hard to believe, even having seen it: *people were not told.*

Imagine it from the employee's side, because nobody at the time did. You come in on an ordinary morning. You log into the system to serve a customer or place an order — the thing you've done capably for years. And the processes you've always followed, the naming conventions you've always used, the vocabulary that made you fluent in your own job — gone. Nobody communicated that things were changing. Nobody communicated what they were changing *to*. Nobody explained what the new terms meant.

One day you were an expert. The next day you couldn't do your job. And no one had thought you were worth telling.

The damage ran all the way downstream, exactly as you'd predict. Customers waiting on employees who were lost in their own tools. Orders fumbled. Data entered wrong — because when people don't know what the new fields and terms mean, they guess, and every guess becomes a record the organization will someday trust. And through all of it, nobody had thought about the people doing the work.

The truth is, having the conversation *prior* would have changed everything.

When I came in, we did things differently. We communicated. We included. And I'll tell you something about where the turnaround actually came from: an amazing systems administrator — not an executive, not a consultant, someone close to the work — single-handedly came up with a new plan for how the alignment could be done. My job, once I was in place, was to put the bells and whistles around that plan and give it the support to succeed. The answer, once again, was already inside the organization. It just needed someone to ask — and the conditions to be heard.

But I won't pretend it was a clean victory. By the time we started doing it right, a lot of bad blood had already moved through the bloodstream. The project had a taste people couldn't unremember. Every conversation started from behind. All of it — every ounce of it — could have been mitigated by two things the earlier cutovers never offered:

Communication. And inclusion.

What Was Actually Happening

Let's name what those silent cutovers really did, because it's more than an oversight.

When you change someone's work without telling them, they don't experience it as a process update. They experience it as theft. Not of their tools — of their *competence*. That employee logging in to a foreign system didn't just lose their naming conventions. They lost, overnight and in front of a waiting customer, the fluency that made them good at their job. And they learned something in that moment that no memo could later unteach:

Decisions about my work happen without me. I find out when it's already done.

That's the lesson silence teaches. Because here's the thing leaders miss: **silence is also a communication.** You cannot opt out of communicating with your people. You can only choose the message. And the message of silence is always the same — *you weren't worth telling.*

Once that message lands, people respond exactly the way you'd respond. They stop extending trust to the next change. They assume the worst version of every rumor — and there are always rumors, because the rumor mill has never once waited for the official announcement. They comply on the surface and resist underneath. And the organization, looking at all this, diagnoses itself with a "change management problem" — as if resistance were a property of the people, rather than a debt run up by the leaders who kept changing their world without a word.

I call it what it is: change debt. Every uncommunicated change borrows against your people's trust. And like all debt, it compounds — which is why, even after we started communicating well, we were still paying interest on cutovers that happened before I ever walked in the door. Good communication couldn't instantly buy back what silence had sold.

That's the part I most want you to hear: **trust damage outlives the fix.** You can repair the system in a quarter. The bloodstream takes years.

Why Before What

Now — this book is about AI, and you may have noticed this story contains none. That's deliberate. The pattern I just described predates AI by decades. But I'm telling it here, at the close of Part One, because this exact pattern is the playbook organizations are running *right now* with AI — at far higher stakes.

Think about what an AI rollout looks like in most companies. An announcement leads with the tool: *"We're launching [platform] next quarter."* The what. Maybe a training date. Maybe a link. And the why — the reason, the intention, what it means for the people receiving the message — is either missing, or buried, or (worse) the unspoken why from Chapter 1 that everyone can smell anyway.

Here's why that sequence fails, every time:

The what tells people's hands what to do. The why tells their hearts whether to care.

When you lead with the what, you hand your people a fact with a hole in the middle of it — and humans do not tolerate holes. They fill them. With the seminar statistics from Chapter 3 fresh in your mind, you already know what they fill them with: the scariest available story. *It's about headcount. It's about replacing us. It's about squeezing more from fewer.* You didn't say that. You didn't have to. You said nothing, and nothing has a sound.

But when you lead with the why — a real one, anchored in the business, safe to say out loud — something different happens. The change gets a *meaning* before it gets a mechanism. People can locate themselves in it. They can disagree with it, even, which is infinitely healthier than fearing it, because disagreement is a conversation and fear is a departure.

One of the deepest beliefs I hold about leadership is this: **people support what they understand.** Not what they're told. Not what they're trained on. What they *understand*. Understanding is built in only one order — why, then what, then how — and no amount of polish on the what can compensate for a missing why.

And communication is not an announcement. It's a process — before, during, and after. Before: here's what's coming, here's why, here's what it means for you, here's what the new words mean. During: here's what's working, here's what we're hearing, here's what we're fixing. After: here's what changed because you spoke up. One memo is not communication. One town hall is not communication. Communication is the drumbeat that makes the rumor mill redundant.

Inclusion Is the Other Half

There's one more lesson sitting in my story, and it's the systems administrator.

The plan that saved that integration didn't come from the leadership team, and it didn't come from me. It came from someone close enough to the work to see what the boardroom couldn't. My contribution was creating the conditions — the platform, the sponsorship, the bells and whistles — for that idea to become the plan of record.

That's what inclusion actually is. Not a courtesy. Not a culture initiative. A *search strategy* — the only reliable way to find the knowledge your organization already holds. The people doing the work know where the process really breaks, what the vocabulary really means, which changes will help, and which will detonate. Communicate *at* them and you get compliance at best. Communicate *with* them and you get their intelligence.

You've now seen this pattern in every chapter of Part One. The organization in the Prologue had the answer and lacked the conditions for it to influence the decision. The reconciliation answer sat in vendor demos for two years. The fear data was one honest meeting away. The colleague with the shadow tool was a walking, measurable business case nobody asked about. And here, the integration plan lived in a systems admin waiting to be heard.

Your organization already knows most of what it needs to know. Communication and inclusion are how leadership finds out.

Closing Part One

So let's gather what Awareness actually means, because you've now done the work of it.

You've examined your why — and made it one you can say out loud (Chapter 1). You've faced the two ways AI fails: rushing without understanding, and freezing without it (Chapter 2). You've named the fear, and learned it's uncertainty wearing a costume (Chapter 3). You've replaced the false contest with the AND (Chapter 4). You've gone looking for the AI already in your building and read it as the requirements document it is (Chapter 5). And now you know the thread that runs through all of it: none of those chapters works in silence. The why must be *communicated*. The fear must be *heard*. The AND must be *said and shown*. The shadow users must be *invited in*.

Awareness, it turns out, isn't something an organization has. It's something leaders create — out loud, on purpose, in the right order.

Which raises the question Part Two exists to answer. Your people now know why. They're beginning to trust the conversation. The conditions for honesty are forming.

So: are you actually ready?

Not "do you feel ready." Not "is the board impatient." Ready — measurably, honestly, dimension by dimension. That's Diagnose. Turn the page.

The Executive Brief

Key Insight

Silence is also a communication — it tells people they weren't worth telling, and that message compounds into change debt that outlives the fix. Lead every change with why before what, because the what tells people's hands what to do and the why tells their hearts whether to care. People support what they understand.

Questions Every Leader Should Ask

- Think of our last significant change. When did the average employee learn about it — before it happened, as it happened, or from a colleague?

- Can the people affected by our current initiatives state, in their own words, why we're doing them?
- Where are we carrying change debt from past rollouts — and are we mistaking that debt for a "resistant culture"?
- Who closest to the work has ideas we've never created the conditions to hear?
- If we announced our AI plans tomorrow, what would lead: the tool, or the reason?

Monday Morning Actions

1. Audit your last major announcement. Count the sentences about the what versus the why. If the why lost, rewrite it as the communication you should have sent — and use that as your template going forward.

2. Establish the drumbeat for your AI initiative now, before launch: a recurring, predictable communication rhythm — before, during, and after — so the official channel outruns the rumor mill.

3. Find your systems administrator. Identify two or three people close to the work your AI plans will touch, and put them in the design conversation this month — not the feedback session after decisions are made. The plan that saves you may be theirs.

Your AI Prompt

"Here is a change announcement I'm preparing: [paste draft]. Rewrite it to lead with why before what. Then critique it from the perspective of a skeptical frontline employee who has lived through poorly communicated changes: what holes will they fill with fear, what questions does it fail to answer, and what does it ask of them without acknowledging what it means for them?"

Reflection

Recall a time in your own career when a change was done to you rather than with you — when you found out after it was decided. What did it cost the people who decided it, in your trust and effort, that they never knew they paid? Now ask: where might you be collecting that same invisible bill from your own people?

Your 90-Day Checkpoint

Part One is complete when three things are true: you can state your why in one sentence your whole leadership team repeats consistently; you've held the fear conversation and can name your people's top concerns in their words; and you have a communication drumbeat scheduled — not planned, scheduled. If all three are true, you've built Awareness. Part Two will test whether you're ready.

PART TWO — DIAGNOSE

"Are we actually ready?"

In Part One, you built Awareness — the honest why, the named fears, the people-AND-machine conviction, the communication drumbeat. That work prepared your organization to have the AI conversation.

Part Two is where you find out whether you're prepared to act on it.

Diagnosis is the discipline most organizations skip, because it feels slow when everyone is impatient and unnecessary when everyone is excited. It is neither. The next four chapters will teach you to find the real problem, measure the right things, examine your leadership honestly, and refuse to automate confusion. Then Chapter 11 will hand you the instrument that pulls it all together: the AI Ready Scorecard™.

Before you spend a dollar, you're going to learn where to aim.

CHAPTER 7

You're Solving the Wrong Problem

Back to the Boardroom

In the Prologue, I told you about a meeting — the one where an organization approaching a billion dollars in revenue celebrated six figures collected by an AI tool, while nothing that created the problem changed.

I brought you back here on purpose. Because in the Prologue, I told you *what happened*. Now that you've done the work of Part One, I want to show you how to *see* it — because the mistake made in that boardroom is the single most common mistake in AI adoption, and unless you learn to catch it, you will make it too. Intelligent, experienced leaders make it every day.

Here's the moment again, slowed down.

The organization had a genuine, painful, expensive problem: accounts receivable. And when the AI suggestion came — *"a tool that could help us with collections"* — every head in the room nodded. Why wouldn't they? Collections was where the pain lived. Collections was where the numbers looked worst. Collections had a department, a budget line, and a report that landed on the executive team's desk every month.

But walk upstream with me, the way I was walking upstream in my head while those heads nodded.

An account doesn't begin in collections. It ends there. Before any invoice ever went to collections, a customer was onboarded — maybe with credit terms nobody scrutinized. A contract was written — maybe with billing language that guaranteed disputes. Work was performed and recorded — maybe in systems that didn't talk to each other. An invoice was generated — maybe inaccurately, because the data feeding it was inconsistent. It went out — maybe late. A dispute came in — maybe to no clear owner. Payment terms slipped — maybe because nobody was watching until the aging report turned red.

By the time an account reached collections, dozens of decisions — spread across sales, operations, finance, and technology, accumulated over years — had already determined whether that invoice would ever be paid on time. Collections wasn't where the problem lived.

Collections was where the problem *arrived*.

So when the organization pointed its shiny new AI at collections, it wasn't solving the problem. It was industrializing the cleanup. And because nothing upstream changed, the pipeline kept manufacturing new bad debt at exactly the old rate — now with a more efficient janitor at the end of it.

That's the mistake. Now let's understand why smart people keep making it.

Why Pain Gets All the Attention

Organizations aim at symptoms instead of sources for reasons that are completely rational in the moment — which is what makes the pattern so durable. It's worth naming them, because you'll recognize every one.

Pain is visible. Causes are quiet. The collections number was on a monthly report in bold. The contract language, the onboarding gaps, the disconnected systems? No report. No red number. Nothing upstream ever *looked* like an emergency — it just quietly created one downstream, month after month, for years.

Pain has an owner. Causes are orphans. Collections had a department you could call into a meeting. The upstream problem belonged to sales *and* operations *and* finance *and* IT — which, in organizational terms, means it belonged to no one. Cross-functional problems are the loneliest problems in business, and they are precisely where the biggest money hides.

Pain has a budget line. Causes need a case. Spending against a visible symptom is easy to justify — the symptom justifies itself every month. Spending against a root cause requires someone to first prove the cause exists, then prove it connects to the pain, then convince four departments to change. Guess which purchase order gets signed first.

And here's the one specific to this book: AI vendors sell to your pain. They have to — it's where your attention already is. Every demo you will ever sit through is aimed at the number that's already keeping you up at night. Nobody demos against a problem you haven't noticed yet. Which means the market itself is organized to help you automate your symptoms, and it will do so enthusiastically, with excellent slideware. The vendor in our story wasn't dishonest. The tool worked! It collected real money. The vendor simply answered the question the organization asked.

The organization asked the wrong question.

Solve where value is created — not where pain becomes visible.

That's the principle at the heart of this chapter, and it extends far beyond receivables. Every organizational result has an origin, and the origin is almost never where the result finally gets your attention. Customer complaints surface in the call center; they're created in product, policy, and process. Turnover surfaces in exit interviews; it's created in management, workload, and recognition, months earlier. Quality failures surface in inspection; they're created in design, training, and inputs. Missed forecasts surface in the quarterly review; they're created in a hundred daily execution gaps nobody was watching — which, if you remember my *Revenue Is Too Late* lesson, is a story we'll finish in the next chapter.

The pattern never changes. Pain is downstream. Value is upstream. And technology pointed downstream doesn't transform anything — it just processes the consequences faster.

The Upstream Walk

So how do you actually find where value is created? You don't need a consulting engagement. You need a discipline I call the upstream walk, and you can do it in a conference room with the people who touch the work.

Take your painful outcome — the number everyone wants to aim AI at — and walk it backward with five questions:

One: Where does this outcome first become inevitable? Not where it becomes visible — where it becomes *locked in*. For our receivables, an invoice's fate was largely sealed at onboarding and contract, months before collections ever saw it. Find the moment of no return; that's usually your real target.

Two: Who touches this before it breaks? List every hand, system, and decision between origin and pain. The length of that list is the size of the territory the boardroom conversation is ignoring.

Three: Where does the data degrade? Follow the information, not just the work. Every handoff where data gets rekeyed, guessed at, or lost is a place your future AI will inherit garbage — and AI fed garbage doesn't fix your problem, it accelerates it. (Never automate confusion; Chapter 10 will hold you to it.)

Four: What would have to be true upstream for this problem to never arrive here? This question flips the room from cleanup thinking to prevention thinking. "What would have to be true for an account to never reach collections?" produces a completely different project list than "how do we collect faster?"

Five: If we fix it here, what's still broken? The killer question — the one nobody asked in that boardroom. If the collections tool works perfectly, invoices are still inaccurate, still late, still disputed, and the bad-payer percentage hasn't moved. If your proposed AI initiative passes this test — if fixing it *there* actually stops the problem from being manufactured — you've found the source. If it doesn't, you've found a symptom wearing a solution's clothes.

Run those five questions before any AI initiative, and something changes in the room: the conversation stops being about the tool and starts being about the business. Which is exactly where it belonged all along.

A Word of Nuance Before You Swing the Pendulum

Now, before you cancel every downstream project in your portfolio — a caution.

Sometimes you *should* treat the symptom. If the patient is bleeding, you stop the bleeding. An organization in a cash crisis may genuinely need collections help this quarter, and there's no shame in that. The discipline isn't "never work downstream."

The discipline is: **know which one you're doing.** Treating the fever while curing the disease is medicine. Treating the fever and *declaring the disease cured* is what happened in that boardroom. If you deploy a downstream tool as a bridge, then say so out loud, fund the upstream fix in the same breath, and measure yourself on the upstream numbers — did the bad-payer percentage move? did invoice accuracy improve? — not the cleanup numbers.

Which brings us to the applause.

Because the deepest failure in that boardroom wasn't the tool selection. It was the celebration. Six figures collected, high-fives all around, "where else can we use AI?" — and not one person asked whether the conditions that created the problem had changed. They had aimed at the wrong target *and then scored themselves on hitting it.*

Solving the wrong problem is a diagnosis error. Celebrating it is a measurement error. And measurement errors are how wrong problems get funded for years.

That's the next chapter.

The Executive Brief

Key Insight

Pain is downstream; value is created upstream. Organizations aim AI where problems become visible — because pain has reports, owners, and budget lines — instead of where problems are manufactured. Technology pointed at symptoms doesn't transform the business; it industrializes the cleanup. Solve where value is created, not where pain becomes visible.

Questions Every Leader Should Ask

- For the problem we most want to aim AI at: where does that outcome first become *inevitable* — not visible, inevitable?
- If our proposed AI initiative works perfectly, what will still be broken?
- Which of our biggest problems are cross-functional orphans — owned by everyone and therefore no one?
- Is the initiative in front of us a cure, or a bridge? Have we said which — out loud — and funded the cure alongside the bridge?
- What question is our vendor's demo answering, and is it the question we should have asked?

Monday Morning Actions

1. Pick the one painful number your organization most wants AI to fix. Schedule ninety minutes with the people who touch that work — not just their managers — and run the five-question upstream walk.

2. Map the moment of no return: identify where in the process the bad outcome gets locked in, and name who owns that moment today. If the answer is "no one," you've found your first real project.

3. Review your active AI initiatives (including anything surfaced by Chapter 5's discovery) and label each one honestly: source, bridge, or symptom. For every bridge, confirm the upstream fix is funded and owned. For every symptom, reopen the conversation.

Your AI Prompt

"I'm going to describe a painful business outcome my organization wants to fix: [describe it]. Walk me upstream: map the likely process steps, decisions, and handoffs that occur before this outcome becomes visible, identify where the outcome most likely becomes inevitable, where data probably degrades, and which upstream functions would need to change for this problem to stop being created. Then critique this plan: [describe your current or proposed solution] — is it aimed at the source, serving as a bridge, or treating a symptom?"

Reflection

Think of the most persistent problem in your organization — the one that survives every initiative aimed at it. Has anyone ever walked it upstream? Or has every solution, for years, been deployed at the place where it hurts? What might its survival be telling you about where you've been aiming?

Your 90-Day Checkpoint

Before the end of this week, you should be able to state, for your top AI candidate initiative: where the problem is created (not where it's visible), what would have to be true upstream for it to stop occurring, and whether your initiative is a source fix, a bridge, or a symptom. If you can't answer yet, don't move to Build — the most expensive thing you can do in the next sixty days is solve the wrong problem efficiently.

CHAPTER 8

You're Measuring the Wrong Things

Several Million Dollars Behind

Early in my career, I worked for a SaaS company alongside three other directors.

We had a new executive leader who was intensely focused on revenue. Every leadership meeting centered on the numbers — revenue, forecasts, financial performance. If you had judged that organization by its meeting agendas, you would have concluded that nobody on earth cared more about revenue than we did.

Then one day, during a third-quarter leadership meeting, that executive asked a question that stopped all four of us in our tracks.

"If things are going so well, why are we several million dollars behind last year's revenue?"

The room went silent.

Not because we knew the answer.

Because none of us even knew we were behind.

Sit with that for a moment, the way I've had to. We were in the third quarter of the year — and this was the first time anyone in that room had been shown that we were significantly off pace. There was no opportunity left to adjust our execution. No opportunity to shift strategy. No opportunity to redirect resources or pivot. The number had finally surfaced, and all it could do was report what had already happened.

And here's the detail that has stayed with me longest: the leader most obsessed with revenue was leading the organization that got blindsided by it. Every meeting had been about the number. And the number still arrived as a surprise.

That's not a contradiction. That's the lesson.

Staring at the scoreboard doesn't move the score. We had spent months reviewing the result of the business while nobody was instrumenting the *work* of the business — the daily execution that was quietly falling behind, week after week, in plain sight of no one.

Revenue is a lagging indicator. It tells you where you've been — not where you're headed. By the time it declines, the decisions, behaviors, and execution gaps that caused the decline have been in place for weeks or months. The meeting where you finally see it isn't a strategy meeting.

It's an autopsy.

What Was Actually Happening

Let me give you the language for what that room was missing, because once you have it, you'll see this everywhere.

Every business runs on two kinds of numbers.

Lagging indicators report outcomes: revenue, profit, churn, collections, turnover. They're real, they matter, and they share one defining trait — by the time they move, you can no longer influence them. They are the past, arriving.

Leading indicators report the work that produces outcomes: Are priorities clear? Are commitments being completed? Are roadblocks being removed quickly? Are customers getting what they need? Are decisions being made fast enough? Are people focused on the work that actually moves the business forward?

I like to say it this way: **revenue tells you the score. Rhythm tells you whether you're winning.**

Your rhythm is the daily execution of the business — and rhythm is where results are manufactured. When your rhythm changes, revenue eventually follows. Always. The only question is whether you noticed the rhythm change months before the revenue did, or whether you found out the way my leadership team found out: in the third quarter, with nothing left to do about it.

Organizations that only watch lagging indicators end up forced into conversations like *"we need to pivot," "we need to cut costs," "we need to restructure."* Those aren't strategies. They're reactions — the vocabulary of leaders who got their first look at the problem after it finished happening.

Revenue matters. But revenue is the report card — not the classroom.

The Boardroom, Revisited One Last Time

Now hold that language up against the story this book opened with, because the collections celebration and my silent Q3 meeting are the same failure wearing opposite costumes.

In my SaaS story, a lagging indicator *blindsided* a leadership team — the bad news arrived too late to act on.

In the collections story, a lagging indicator *flattered* a leadership team — six figures collected, applause, high-fives — while every number that actually mattered went unexamined. Did the percentage of habitual late payers move? Did invoice accuracy improve? Were invoices going out on time? Were fewer accounts reaching collections at all?

Nobody knew. Worse — nobody asked. The organization measured the cleanup and mistook it for the cure.

That's the danger of lagging indicators: they don't just arrive too late to help you. They will happily tell you a success story while the disease progresses. Chapter 7 said it and now you can see the machinery of it: solving the wrong problem is a diagnosis error, but *celebrating* it is a measurement error — and measurement errors are how wrong problems get funded for years. A bad initiative with good-looking metrics is nearly immortal.

If you take one sentence from these two stories combined, take this one:

Measure what creates the outcome — not just the outcome itself.

What This Means for Your AI Initiative

Here's why this chapter sits in the Diagnose section of an AI book: the AI industry is a lagging-indicator factory, and most organizations walk straight into it.

Look at how AI success gets reported in the average company. Licenses purchased. Users enrolled. Training sessions completed. Prompts run this month. A dashboard full of adoption theater — numbers that are easy to collect, pleasant to present, and tell you almost nothing about whether AI is changing how the business performs. They are the six figures collected. They photograph well while nothing upstream changes.

If you want to know whether your AI initiative is actually working — months before the business results arrive — watch the rhythm instead:

Are leaders visibly using the tools themselves — or just sponsoring them from a distance? Your people watch feet, not memos.

Is it safe to experiment? Are people asking beginner questions in the open, or is usage quiet and defensive? (You built this in Chapter 3; here's where it becomes measurable.)

Are use cases moving between people and teams? One person saving an hour is a trick. That trick spreading to twelve people is adoption.

Is saved time being reinvested — and can anyone say into what? A day a week, like my colleague in Chapter 5 was saving, only matters if you know where it went.

Are barriers being removed quickly? Track the age of the oldest unresolved blocker. It measures your leadership's response speed better than any survey.

Are employees proposing new uses? The moment ideas start flowing uphill, your initiative has crossed from compliance into belief.

Those are leading indicators of AI success — and notice that not one of them requires a data science team. They require leaders paying attention to the classroom instead of waiting for the report card.

Becoming Measurement-Ready

So here's the discipline, and it belongs *before* implementation — that's why we're doing it in Diagnose, before you've spent anything. Four commitments, on paper, before any tool is purchased:

Baseline. Record today's numbers for the problem you chose in Chapter 7 — at the source, not just the symptom. If the problem is receivables, that means invoice accuracy, on-time invoicing, dispute rates, and bad-payer percentage — not just the collections balance. You cannot claim improvement later against a baseline you never took.

Both scoreboards. Define your lagging indicators (the business outcomes that should eventually move) *and* your leading indicators (the rhythm signals that should move first). If you can't name what should change in the first sixty days, you don't have a measurement plan — you have a hope with a dashboard.

Cadence. Decide now who reviews these numbers, how often, and in which meeting. Leading indicators reviewed quarterly are just slow lagging indicators. Rhythm gets reviewed at the speed of the work — weekly, in most cases.

Pivot triggers. Write down, in advance, what result would make you change course — and what you'd do. This is the step that would have saved both boardrooms in this chapter. It converts measurement from scorekeeping into steering, and it makes it politically possible to say "this isn't working" later, because you agreed on the tripwire before anyone's reputation was attached to the initiative.

Do this and something quietly powerful happens: your organization becomes capable of an honest answer to "is the AI working?" — which, as we saw in the Prologue, is a question most organizations cannot actually answer. They can only tell you whether the demo of success is still running.

Lead the work. The numbers will follow.

And that raises the next diagnostic question. A measurement plan needs people with the authority, visibility, and safety to act on what it reveals — because a leading indicator nobody is empowered to respond to is just early bad news. Do you have those people? Are *you* one of them?

That's Chapter 9.

The Executive Brief

Key Insight

Lagging indicators report what already happened; by the time they move, you can't influence them — they arrive as autopsies or, worse, as false celebrations. Revenue tells you the score; rhythm tells you whether you're winning. Measure what creates the outcome, not just the outcome itself — and define both scoreboards before your AI initiative begins.

Questions Every Leader Should Ask

- What do our leadership meetings actually review — the work of the business, or the results of the business?
- For our current priorities, could a bad quarter still surprise us in month nine? What would have warned us in month two?
- If our AI initiative were quietly failing right now, which number would tell us first — and does anyone review it?
- Are our AI metrics measuring business change, or adoption theater — licenses, logins, and training completions?
- What result would make us change course, and did we define it before or after we got attached to the initiative?

Monday Morning Actions

1. Audit your last three leadership meeting agendas. Mark every item as leading (the work) or lagging (the results). If lagging wins, redesign the standing agenda so rhythm is reviewed before financials.

2. For the problem you chose in Chapter 7's upstream walk, capture the baseline this week — at the source, not the symptom. Date it. This becomes the "before" photo your entire AI initiative will be judged against.

3. Draft your two scoreboards for that initiative: three to five leading indicators, two or three lagging ones, an owner and cadence for each, and one written pivot trigger. One page, before any vendor conversation.

Your AI Prompt

"I'm preparing to measure an AI initiative aimed at this business problem: [describe it]. Help me build a measurement plan: (1) the baseline metrics I should capture now, at the source of the problem rather than its symptom; (2) leading indicators that would show within 30–60 days whether the initiative is taking hold; (3) lagging indicators that prove business impact; (4) a review cadence and owner for each; and (5) two or three pivot triggers — specific results that should make us change course. Challenge any metric that measures activity rather than outcomes."

Reflection

Think of the worst business surprise of your career — the number that arrived too late to act on. Now trace it backward: what was observable three months earlier, and why wasn't anyone looking there? What's the equivalent blind spot in your organization today?

Your 90-Day Checkpoint

Before you leave Diagnose, your top AI initiative should have a one-page measurement plan: a dated baseline at the source, leading and lagging indicators with owners and cadence, and pre-agreed pivot triggers. If you can't produce that page, you're not measurement-ready — and an initiative you can't measure is an initiative you can only celebrate.

CHAPTER 9
Do You Have the Right Leaders?

The Seat I Was Sitting In

I need to take you back to that boardroom one last time — but this time, I'm not going to show you the room.

I'm going to show you my seat.

You know the story by now. The receivables crisis. The AI tool aimed at collections. The applause, weeks later, for six figures recovered while the machinery that manufactured the problem ran on untouched. In Chapter 7, I showed you the diagnosis error. In Chapter 8, the measurement error. But there's a third layer to that story, and it's the one that took me the longest to talk about — because I was in it.

I didn't walk into that meeting unprepared. I walked in carrying the benchmarks. The aging trends. The invoice accuracy data. The operational evidence that showed, plainly, where the breakdowns were happening and how long they had been happening. I was the person in the organization whose job was transformation — which means I was the person whose job was to see exactly what I saw.

And the room went a different direction anyway.

Not because anyone was unintelligent. Not because anyone acted in bad faith. The vendor suggestion was concrete, exciting, and easy to say yes to. My evidence pointed somewhere harder — upstream, into cross-functional territory, into years of accumulated decisions. The concrete thing won. The applause followed.

Here's the part that stays with me, though. After the celebration, nobody came back.

Nobody came back to the person who had raised the concern and asked: did we make the mark? Did the bad-payer percentage move? Did invoice accuracy improve? All the benchmarking that would have answered those questions was sitting right there — I had brought it into the room myself. It was never requested again.

And that's how I knew, with certainty, that they would be back in that same room within no time at all. Because nothing upstream had changed — including the thing furthest upstream of all: the conditions around the person responsible for changing it.

What Was Actually Happening

For a long time, I told that story to myself the easy way: *they didn't listen.* But the easy version is the useless version, because it makes the lesson about individuals — and the lesson was never about individuals.

Here's the harder, truer version.

That organization employed a transformation leader. It gave that leader accountability for outcomes. What it never fully gave that leader were the conditions the accountability required:

Sponsorship — an executive voice that said, before the room decided anything, "the transformation lead has evidence we need to weigh." Not agreement. Weight.

A platform — a standing place in the decision process where upstream evidence got examined *before* solutions got selected, instead of a seat from which to watch the selection happen.

Visibility — enough cross-functional reach to connect the dots between sales, operations, finance, and technology, in an organization where each function saw only its own segment of the pipeline.

Cultural safety — the established, demonstrated norm that challenging an exciting idea in that room was a contribution, not a career risk.

Acknowledgment — the follow-up loop. The organizational habit of returning to the person who flagged the problem and asking whether it was actually solved.

Take away those five things, and it doesn't matter how good the evidence is or how capable the person carrying it is.

The evidence attends the meeting. It just doesn't influence the decision.

And I want to be fair here, because fairness is the point. I've asked myself the accountability question too. Could I have built the coalition earlier — pre-sold the upstream case one executive at a time before it ever reached the room? Could I have made the benchmarks harder to look away from? Probably, in part. I've gotten better at exactly those moves in the years since, and I'll teach them in this book.

But hear the deeper truth underneath my self-critique: an organization whose only safeguard against a multi-million-dollar aiming error is *one person's persuasiveness on one afternoon* does not have a leadership system. It has a lottery.

Which brings me to the sentence this entire chapter exists to deliver:

Organizations don't fail because they lack capable people. They fail when capable people are expected to deliver transformation without the conditions that make transformation possible.

The organization had all the information it needed. What it lacked were the conditions for that information to change the decision.

Why This Is a Diagnostic Chapter

You might be wondering why a chapter about leadership sits here, in Diagnose, between measurement and process — rather than somewhere in a chapter about "change management."

Because leadership readiness is not a soft topic. It is a hard, checkable precondition — and it fails silently. Your data readiness will announce its failures loudly; your tools will crash; your budgets will overrun visibly. But a transformation leader without sponsorship fails invisibly. Meetings still happen. Reports still circulate. Everyone is busy. And the organization aims at the wrong target with full confidence, because the person positioned to catch the error had no mechanism by which catching it would matter.

So before you spend, you diagnose three layers of leadership — honestly, the way you'd audit a balance sheet.

Layer one: the executive team. Is your leadership team actually aligned on the why from Chapter 1 — or politely assembled around a vague one? Alignment is checkable: ask each executive to write the one-sentence why independently, and read the answers side by side. Then check sponsorship. Sponsorship is not approval; approval signs the budget, sponsorship shows up. A sponsored initiative has an executive who attends the working sessions, uses the tools personally (Chapter 8's first leading indicator), spends political capital when functions collide, and says the why out loud, repeatedly, in their own words. If your initiative has approvers but no sponsor, it is already in trouble — it just doesn't know it yet.

Layer two: the transformation leader. Someone in your organization is, or will be, the person accountable for making this change real. Now audit their conditions against the five I paid for the hard way. Do they have the *authority* to influence decisions, or only the responsibility to coordinate them? Do they have *access* — a direct line to the executive team, or three layers of translation? Do they have *visibility* across every function the problem touches? Do they have *cultural safety* to challenge the room's favorite idea — and has anyone ever actually seen them do it and be thanked? And is there an *acknowledgment loop* — a standing expectation that the organization returns to their flagged concerns and closes them out with evidence?

Here's the test I give executive teams, and I'd rather you sit with it than rush past it: **think of the last time your transformation leader (or any leader) challenged a direction this team was excited about — and the challenge changed the outcome.** If you can name a recent, specific instance, you may have the conditions. If you can't, then one of two things is true: either every decision you've made recently was right, or the challenges have stopped reaching you. One of those is very unlikely. The other is very expensive.

Layer three: ownership. Beyond the sponsor and the transformation leader, every AI initiative needs named owners for the unglamorous middle: who owns adoption? Who owns communication — the drumbeat from Chapter 6? Who owns removing the barriers Chapter 8 told you to track? Who owns the decision when two functions disagree about a workflow? Write the names down. If any answer is a committee, translate that honestly: a committee owning something means everyone can explain why it wasn't them. If everyone owns it, no one owns it.

The CIO Question

By now, some of you are asking the practical question: *we have a CIO — can't they lead this?*

Maybe. But before you answer, let me tell you about a road.

I worked inside a company where the CIO was genuinely excited about transformation. This is not a story about a reluctant executive dragged into change — the desire was real. This CIO kicked off a transformation initiative with one of the Big Four, and on paper, everything a board wants to see was present: an engaged technology leader, a marquee consulting partner, an initiative with a name.

It was doomed before it started. Not because of who the leader was — because of what never happened.

There was never a defining moment for what transformation actually meant at that company: the level of resources it would take, and the level of commitment the organization would have to make for it to be real. No one wrote that sentence down, so no one had to fund it, staff it, or stand behind it. And there were no clear metrics for what success would look like — in fact, metrics made this leader uncomfortable. I understand why. When no one has defined what good looks like, a metric doesn't feel like an instrument; it feels like a trap. That discomfort wasn't a character flaw. It was a symptom of the missing definition — the exact failure Chapter 8 warned you about, arriving on schedule.

So here's what the initiative became. Instead of looking at an entire process, we looked at pieces and parts of the process. And I believe I know why: in this leader's mind, something was better than nothing. That instinct is human, and in an operator's world, it's usually right — you patch what you can reach.

But when an entire system is broken, fixing a few steps inside it doesn't fix what's wrong. It magnifies it. Sometimes it makes the issue even more broken.

You've driven this road. When potholes appear, the city fills them — one here, one there, each patch a small victory. But keep patching long enough and the road itself becomes the problem: bumpy, uneven, a surface of mismatched fixes. Now it's not the holes damaging your car — it's the road. The shocks wear, the tires pop, and every patch made the whole slightly worse. At some point there is only one honest fix: close the road, rebuild the entire surface, smooth it, repaint the lines, and reopen it better than it was.

An operator fills potholes. A transformation executive closes the road.

That's the distinction, and it has nothing to do with talent. A transformation executive walks into that same broken system and says: *all of this must be done. Here are your options for doing it. This is what good looks like. And this is how one initiative moves the entire organization* — not fragments pulled from a broken system and polished while the system stays broken.

The CIO role, done well, is built to operate: uptime, security, stability, cost. Those instincts — protect the system, minimize disruption, patch what you can reach — are exactly what you hire for, and exactly what makes a great CIO great. Transformation asks for opposite instincts: define the destination before touching anything, demand the full commitment or decline the work, hold the whole system in view, and welcome the metrics — because the metrics are how you'll know the road actually got rebuilt.

Some CIOs carry both sets of instincts. Genuinely. If yours does, you may not need a second seat — you need to give that one leader both mandates explicitly, with the five conditions attached to each.

But if yours doesn't — and this is where I need you to be honest with yourself, the way this entire chapter has asked you to be — then handing your CIO the transformation mandate isn't a promotion. It's a miscasting. You are asking a capable operator to close the road when everything in their training, their incentives, and their instincts says *keep the traffic moving*. And when the fragmented fixes multiply and the initiative stalls, the organization will quietly conclude the CIO failed — when the truth is the organization never defined the work, never committed the resources, never set the measures, and never gave anyone the authority to close the road.

That's not a failed leader. That's a leader set up to fail. And you'll have lost two things at once: the transformation, and possibly a very good CIO.

So before you assign the seat, ask three questions, in this order. *Has this organization defined what transformation means here — the resources, the commitment, the measures — in writing?* If not, no one can lead it yet, whatever their title. *Does the leader we're considering have the transformation instincts — will they tell us "all of it must be done" even when we want to hear "we can start small"?* And *are we prepared to give whoever takes the seat all five conditions from this chapter — including the authority to close the road?*

If the answer to any of these is no, the problem isn't your CIO. It's the conditions. Which is precisely where this chapter began.

If You're the Leader Being Asked

Before I close this chapter, I want to speak directly to one reader for a moment — because some of you holding this book aren't the CEO. You're me, in that seat. You're the person about to be handed accountability for an AI transformation.

Negotiate your conditions *before* you accept the accountability. Not as a power grab — as due diligence, for their sake as much as yours. Ask for the sponsor by name. Ask what happens when your evidence conflicts with the room's enthusiasm — and ask for a live example of the last time it did. Ask how your concerns get closed out. If the answers are vague, that vagueness is information: you are being offered a lottery ticket, not a role.

There's no shame in taking the seat anyway — sometimes you build the conditions from inside, the way I've since learned to. But take it with your eyes open, because you now know exactly what the seat costs when the conditions are missing. I've paid that bill. Consider this chapter the receipt.

One More Thing the Right Leaders Do

Leadership readiness, diagnosed honestly, protects you from the failure pattern of this entire Part: the wrong problem, confidently funded (Chapter 7), measured by flattering numbers (Chapter 8), unchallenged by the person who saw it coming (this chapter).

But there's one more thing the right leaders refuse to do, even with the right problem, the right measures, and the right conditions. They refuse to point powerful technology at a process nobody understands — because automating confusion doesn't resolve it. It multiplies it.

That's Chapter 10.

The Executive Brief

Key Insight

Organizations don't fail for lack of capable people; they fail when capable people carry accountability for transformation without the conditions it requires — sponsorship, platform, visibility, cultural safety, and an acknowledgment loop. Evidence can attend every meeting and still influence nothing. Conditions, not intelligence, determine whether your organization can catch its own aiming errors.

Questions Every Leader Should Ask

- Who is accountable for making our AI initiative real — and do they have authority, access, visibility, safety, and follow-up, or just responsibility?
- Does our initiative have a sponsor who shows up, or only approvers who signed off?
- When did someone last challenge a direction this executive team was excited about — and change the outcome? Can I name the specific instance?
- When someone flags a problem here, does the organization ever come back to them with evidence it was solved?
- Written down, who owns adoption, communication, barrier removal, and cross-functional disputes? Are any of those names actually a committee?
- If we handed our CIO the transformation mandate tomorrow, would we be giving them a defined, resourced, measured commission — or a title and our hopes?

Monday Morning Actions

1. Run the alignment check: every executive independently writes the one-sentence why. Read them aloud together. Gaps found here cost nothing; gaps found mid-implementation cost everything.

2. Audit your transformation leader's five conditions with them in the room — sponsorship, platform, visibility, safety, acknowledgment — and ask them to score each one candidly. Then fix the lowest score this month. (If their candid scores are all high and you suspect politeness, that's your cultural safety score.)

3. Create the acknowledgment loop: a standing agenda item where previously raised concerns are revisited with evidence — moved, resolved, or still open. Assign it an owner. This single habit would have changed the ending of every story in this book so far.

Your AI Prompt

"Act as an organizational design advisor. I'm structuring leadership for an AI initiative. Here's our setup: [describe the sponsor, the transformation/initiative leader and where they sit, and who owns adoption, communication, and barriers]. Stress-test it: where does accountability exceed authority? Where would inconvenient evidence get filtered before reaching decision-makers? What's missing for the initiative leader to successfully challenge a popular but wrong direction? Give me the three highest-risk gaps and a specific fix for each."

Reflection

Somewhere in your organization, there may be a person who already sees your next expensive mistake — who has the benchmarks, the trends, the evidence. Ask yourself honestly: do the conditions exist for that person to change your mind? Or will you find out what they knew the way most organizations do — afterward?

Your 90-Day Checkpoint

Before you leave Diagnose, you should be able to name, in writing: your initiative's active sponsor (not approver), your transformation leader and their score on the five conditions, and the single named owner for adoption, communication, and barrier removal. If any line is blank or reads "committee," your leadership readiness — not your technology — is your first project.

CHAPTER 10
Never Automate Confusion

The Same Engagement, One Layer Deeper

In Chapter 6, I told you about the integration — the organization that had grown by acquisition, the silent cutovers, the morning people logged in and found their vocabulary gone. I told that story for its communication lesson.

Now I need to take you back into the same engagement, because sitting underneath the communication failure was a second discovery — and it's the reason this chapter exists.

My charge, remember, wasn't just to put everyone on one system. It was to get everyone into one *language*. And the deeper we went, the clearer it became why that work had to come first — before any consolidation, before any automation, before anything intelligent could ever be layered on top.

Here's what the landscape actually looked like. Multiple acquired companies, each with its own dialect. The same term meaning two entirely different processes in two different business units. Two different terms meaning the same process. Fields in the system that one company used religiously and another ignored. Codes that meant one thing in one legacy system and something else in the next.

And then the human layer on top of it: people doing their best inside that fog. When employees didn't know what a field meant — especially after cutovers nobody explained — they guessed. And as I said in Chapter 6, every guess becomes a record the organization will someday trust. Multiply that by years, by thousands of transactions, by every acquired company's habits, and you don't have a data problem.

You have an organization that does not agree with itself about what its own words mean.

Now here's the thought experiment I want you to sit with, because it's the whole chapter: imagine that organization — before the language work, before the alignment — deciding to deploy AI. Imagine pointing a system that learns from your data at data built on guesses. Imagine automating workflows across business units that use the same word for different work. Imagine an algorithm confidently routing, matching, reporting, and predicting on top of records that mean different things depending on who typed them.

Nothing about the technology would fail. Every output would arrive on time, formatted beautifully, delivered with total confidence.

And it would be industrialized nonsense.

That's why the language came first. Not because it was glamorous — it was the least glamorous work imaginable. Because clean data depends on shared meaning, and everything you will ever automate depends on clean data. We weren't just aligning vocabulary.

We were making the organization safe to automate.

What Was Actually Happening

Here's the uncomfortable question: how does an organization operate for *years* with that much confusion and not collapse?

The answer is the most important insight in this chapter: **people patch confusion.** Quietly. Daily. Invisibly.

In every confused process, there are employees absorbing the chaos so the customer never sees it. The veteran who knows that "that code means something different when it comes from the west region." The coordinator who re-keys the order because they know the field mapping is wrong. The analyst who mentally adjusts the report because they know which business unit's numbers can't be compared. None of that is documented. All of it is essential. Your confused process *works* — barely, expensively, exhaustingly — because human judgment is duct-taping it together in a hundred places nobody has ever mapped.

That's why confusion is invisible from the boardroom. The process looks fine on paper. The outputs mostly arrive. What you can't see is how much daily human intelligence is being spent making a broken thing look whole.

Now watch what automation does to that arrangement.

Automation removes the patch layer. That's its entire value proposition — take the humans out of the repetitive middle. But in a confused process, the repetitive middle is *where the repairs live*. The veteran's translation, the coordinator's re-key, the analyst's adjustment — gone. The workarounds don't get automated, because nobody knew they existed. Only the broken official process gets automated, and now it runs at machine speed.

A confused process run by people fails slowly, with apologies — a person hits something that doesn't make sense, pauses, asks, fixes. A confused process run by machines fails at scale, with confidence. AI does not pause at nonsense. It has no instinct that says *wait, this can't be right*. It learns your data — guesses and all — and executes what it learned, faster than any human error ever propagated, wearing the credibility of a system that leadership just spent real money on.

This is the book's thesis (AI will never outperform the conditions of the organization that implements it) at its most mechanical. Point technology at a healthy process and you get speed. Point it at confusion and you don't clean up the mess.

You schedule it.

Never automate confusion. Create clarity first.

The Confusion Test

So how do you know? A process that's about to be automated looks documented, looks stable, looks ready — the confusion, as we've established, is hidden inside people. Here's the test I use, five questions, asked of the people who *do* the work, not the people who manage it. (You learned in Chapter 5 who actually knows.)

One: Can two people who do this work describe it the same way? Separately, ask two employees in different teams or units to walk you through the process step by step. If you get two different processes, you don't have one process with variations. You have multiple processes wearing one name — and an automation project that will pick one and break the others.

Two: Does the documented process match the actual one? Take the official process document into the room and ask: "Where does reality differ from this?" The length of that list is your confusion inventory. Automating the document while the organization runs on the reality is how systems get deployed and quietly routed around.

Three: Where do people routinely fix things by hand? Every re-key, every manual adjustment, every "I always double-check that field" is a flag marking a spot where the process is broken and a human is compensating. These are exactly the repairs automation will remove. Find them *before* the automation does.

Four: Do the words mean one thing? Pull ten records and ask people from different units what the key fields mean. In my integration, this single question exposed years of divergence. If the same term carries different meanings, no algorithm can learn your business correctly — it will learn all the meanings at once and blend them into something that is precisely nobody's business.

Five: Where does judgment quietly enter? Ask: "When does this process require you to just *know* something?" Those judgment points are either genuine human work that must stay human (Chapter 4's AND), or confusion so old it's been promoted to expertise. Distinguish them, because automating past the first kind breaks quality, and automating past the second kind hides the breakage forever.

If a process can't be explained consistently by the people who run it, it cannot be automated safely. Not because the technology isn't capable — because you'd be asking it to be certain about something your own organization isn't.

Clarity First Is Not Perfection First

Now, a guardrail — because I can hear a familiar voice using this chapter as a shield.

"We can't automate anything until we fix our data. We can't touch AI until our processes are perfect." Two years later: still vetting, still meeting, still talking. That's the Chapter 2 failure wearing Chapter 10's clothes, and I won't hand paralysis a better costume.

So let me be precise about the discipline. Clarity-first is *scoped*. You don't need a clear organization to automate one process. You need a clear *process*. Run the five questions on the specific workflow you selected in Chapter 7 — the one aimed at where value is created. If it passes, automate it, even while the rest of the company stays gloriously confused. If it fails, then the confusion you found *is* your project's first phase — budgeted, scheduled, and owned, inside the same initiative. Not a reason to stop. A reason to sequence.

And here's the twist that makes this genuinely practical: AI itself is one of the best clarity tools you'll ever have. Before automating anything, you can use AI to *find* the confusion — feed it your process documentation and your people's actual descriptions and let it surface the contradictions; let it profile your data and flag where the same field behaves differently across units; let it compare how the "same" process runs in two regions. In my integration, that discovery work took months of manual effort. Today I'd have a first map in weeks. Using AI to understand your mess is one of its highest-value first assignments.

Just never confuse the mapmaking with the automation. One earns the other.

The End of Diagnose

That closes the fourth diagnostic lens, so look at what you're now holding.

From Chapter 7: the right problem — sourced upstream, where value is created, not where pain surfaces. From Chapter 8: the right measures — baselined at the source, both scoreboards, pivot triggers written in advance. From Chapter 9: the right leaders — a real sponsor, a conditioned transformation leader, named owners. And now from Chapter 10: a clear process — one your own people can describe consistently, safe to hand to a machine.

Four lenses. Any one of them, missing, is how the stories in this book happened. The boardroom in the Prologue failed three of the four at once.

What you need now is a way to hold all of them up to your organization at the same time — honestly, dimension by dimension, with scores instead of impressions. An instrument.

Turn the page. It's time to find out if you're actually ready.

The Executive Brief

Key Insight

Confused processes survive because people quietly patch them — and automation removes exactly that patch layer, executing the broken official process at machine speed with machine confidence. A confused process run by people fails slowly, with apologies; run by machines, it fails at scale. Never automate confusion. Create clarity first — scoped to the process at hand, never as an excuse for paralysis.

Questions Every Leader Should Ask

- For the process we intend to automate: can two people who do the work describe it the same way?
- Where does our documented process differ from the real one — and who is quietly compensating for the gap by hand?
- Do our key terms and fields mean the same thing in every unit that touches this process — or are we about to teach an algorithm several meanings at once?

- Which judgment points in this work are genuine human expertise, and which are old confusion wearing expertise's clothes?
- Are we using "our data isn't ready" as a sequencing plan — or as a permission slip to keep talking?

Monday Morning Actions

1. Run the five-question confusion test on your Chapter 7 process — with the people who do the work, not their managers. Budget ninety minutes. Write down every workaround, guess, and translation you find; that list is your pre-automation punch list.

2. Pull ten real records from that process and ask three people from different teams to interpret the key fields. Score the agreement. Anything below unanimous is a data-meaning project that belongs inside your initiative's first phase.

3. Give AI its first assignment: use it to map the confusion — compare official documentation against your people's actual descriptions, profile the data for inconsistent field usage, and surface the contradictions. Clarity is a deliverable. Schedule it like one.

Your AI Prompt

"I'm evaluating whether a business process is clear enough to automate safely. Here is our official process documentation: [paste]. And here are descriptions of how the work actually gets done, from the people who do it: [paste notes or transcripts]. Compare them: identify every contradiction, undocumented workaround, ambiguous term, and point where human judgment silently patches a gap. Then classify each judgment point as (a) genuine expertise that should remain human or (b) compensation for a process or data defect that should be fixed before automation. Output a prioritized pre-automation punch list."

Reflection

Somewhere in your organization is a veteran whose entire value, on paper, looks like routine work — and whose actual value is silently translating, correcting, and patching a process nobody realizes is broken. What happens to your operation on the day a system replaces the routine and nobody replaced the repairs? Do you know who your patch people are?

Your 90-Day Checkpoint

Before you leave Diagnose, your chosen process should have passed the confusion test — or have a scoped, owned clarity phase built into the initiative plan. If you can state your problem's source, your two scoreboards, your five leadership conditions, and your process's confusion score, you are ready for the instrument that puts it all together. That's next.

CHAPTER 11
The AI Ready Scorecard™

The Assessment Nobody Took

Before we begin, I want to run one last thought experiment on the organization from the Prologue.

Imagine that boardroom — the receivables crisis, the vendor suggestion, the nodding heads — but with one difference. Imagine that before anyone said yes, someone had slid a one-page assessment across the table and said: *before we spend, let's score ourselves honestly.*

Strategic readiness: had they defined the problem at its source, or its symptom? We know the answer — the problem was defined by where it hurt. Measurement readiness: was there a baseline, a scoreboard, a pivot trigger? There was applause instead. Leadership readiness: did the person accountable for transformation have sponsorship, platform, safety, and follow-up? I can tell you personally that the answer was no. Awareness: had anyone communicated a why the workforce could believe? There wasn't one to communicate.

That organization would not have scored well. And here's the point: *discovering that would have cost them a ninety-minute meeting.* Instead, they discovered it the expensive way — after the contract, after the celebration, after nothing changed.

Every story in this book is, in one way or another, an assessment nobody took.

This chapter is that assessment. You've now done the work of ten chapters — examined your why, named the fears, found your shadow users, walked upstream, built scoreboards, audited your leaders, tested for confusion. The Scorecard pulls all of it into a single instrument, because impressions don't survive contact with a boardroom. Scores do.

One warning before you pick up a pen: this instrument only works at one setting — honest. Score what your organization *demonstrates,* not what it intends. Score what you could prove to a skeptical outsider with evidence from the last ninety days. An inflated scorecard isn't optimism. It's the Prologue, with better paperwork.

How the Scorecard Works

The Scorecard measures nine dimensions of readiness. Each dimension has five statements. Rate each statement from 1 (strongly disagree) to 5 (strongly agree), giving every dimension a score between 5 and 25.

Three rules make the results worth having:

Score independently, then compare. Don't fill this out as a group — group scoring produces the politely blended answer. Have every member of your executive team score it alone, then put the results side by side. The *variance* between scorers is some of the best data you'll get: if the CEO scores Cultural Readiness a 22 and the frontline-facing leaders score it a 12, that gap is itself a finding — and you learned in Chapter 9 what it's a finding about.

Evidence, not aspiration. For any statement you rate 4 or 5, you should be able to name the specific evidence — the meeting that happened, the document that exists, the decision that proves it. If the evidence is "we're planning to," the score is a 2.

Low scores are the deliverable. You are not being graded; you're being aimed. A low score isn't a failure — it's the cheapest possible discovery of where your AI investment would have leaked. Celebrate finding it now.

The Nine Dimensions

Dimension 1: Awareness Readiness

Do our people understand why AI matters — and is it safe to talk about?

1. Our executive team has communicated a why for AI that is anchored in our business and safe to say to our entire workforce. *(Chapter 1)*

2. We have held honest conversations about AI fears, and I can name our people's top concerns in their own words. *(Chapter 3)*

3. Our people hear about changes from leadership before they hear about them from the rumor mill. *(Chapter 6)*

4. We know where AI is already in use in our organization — sanctioned and unsanctioned. *(Chapter 5)*

5. A recurring communication drumbeat about our AI plans is scheduled — before, during, and after. *(Chapter 6)*

Maturity indicator: Curiosity is replacing fear, out loud.

If you scored low: Chapters 1, 3, 5, and 6.

Dimension 2: Strategic Readiness

Are we solving the right problem?

1. We have defined the specific business problem our AI initiative will address — in one sentence, consistently, across the executive team. *(Chapter 1)*

2. We have walked that problem upstream and can state where the outcome is created, not just where it becomes visible. *(Chapter 7)*

3. If our proposed initiative works perfectly, we can articulate what will still be broken — and we've labeled the initiative honestly: source, bridge, or symptom. *(Chapter 7)*

4. Our initiative was selected by business value, not by which vendor demo matched our loudest pain. *(Chapter 7)*

5. We can say why this problem, why AI, and why now — without referencing what competitors are doing. *(Chapter 1)*

Maturity indicator: Strategy is choosing the technology — not the other way around.

If you scored low: Chapters 1 and 7.

Dimension 3: Cultural Readiness

Will our culture support adoption?

1. It is demonstrably safe to be a beginner here — people ask basic AI questions in the open, not in private. *(Chapters 3, 4)*

2. Employees can admit to using unsanctioned tools without fear of punishment, and some have. *(Chapter 5)*

3. Our people would describe our AI story as "people AND the machine" — and could point to leadership behavior that proves it. *(Chapter 4)*

4. When change comes, our people extend trust — we are not still paying interest on past rollouts done in silence. *(Chapter 6)*

5. Someone recently challenged a direction leadership was excited about, and the challenge changed the outcome. *(Chapter 9)*

Maturity indicator: People are willing to experiment — and to disagree.

If you scored low: Chapters 3, 4, 6, and 9.

Dimension 4: Leadership Readiness

Do we have the leadership this transformation requires?

1. Our initiative has an active sponsor who shows up, uses the tools, and spends political capital — not just approvers who signed off. *(Chapter 9)*

2. Our transformation leader has all five conditions: sponsorship, platform, visibility, cultural safety, and an acknowledgment loop. *(Chapter 9)*

3. Adoption, communication, and barrier removal each have a single named owner — none of them a committee. *(Chapter 9)*

4. Every executive can state the initiative's why in one sentence, and the sentences match. *(Chapters 1, 9)*

5. When someone flags a problem, our organization returns to them with evidence of whether it was solved. *(Chapter 9)*

Maturity indicator: Leadership is enabling the transformation — not just approving it.

If you scored low: Chapter 9 — and be honest.

Dimension 5: Operational Readiness

Is the process we're about to touch actually ready?

1. The process we intend to automate passes the confusion test: people who do the work describe it consistently. *(Chapter 10)*

2. We have inventoried the workarounds — the manual fixes, re-keys, and translations quietly patching this process — before automation removes them. *(Chapter 10)*

3. The documented process matches the real one, or we know exactly where it doesn't. *(Chapter 10)*

4. We are automating at or near the source of the problem, not industrializing the cleanup. *(Chapter 7)*

5. We know which judgment points in this work must remain human, task by task. *(Chapter 4)*

Maturity indicator: We are improving healthy work — not hiding unhealthy work.

If you scored low: Chapters 7 and 10.

Dimension 6: Data & Technology Readiness

Can our data and systems support what we're asking of them?

1. In the process we're targeting, key terms and fields mean one thing — across every unit that touches them. *(Chapter 10)*

2. We know where our data degrades — every handoff where information gets re-keyed, guessed at, or lost. *(Chapters 7, 10)*

3. The data this AI will learn from is reliable enough that we would defend decisions made on it today.

4. Our systems can integrate with what we're planning to deploy, and someone technical has verified that — recently.

5. Security, governance, and data protection have been evaluated for this specific use — including where our data goes and what the vendor's terms actually say. *(Chapter 5)*

Maturity indicator: Technology is an enabler — not a constraint we discover mid-project.

If you scored low: Chapter 10 — and give AI the mapmaking assignment before the automation one.

Dimension 7: Measurement Readiness

Will we know if we're succeeding — in time to act?

1. We have captured a dated baseline at the source of the problem — not just the symptom. *(Chapter 8)*

2. We have both scoreboards defined: leading indicators that should move in the first sixty days, and lagging indicators that prove business impact. *(Chapter 8)*

3. Each number has an owner and a review cadence, and rhythm gets reviewed at the speed of the work — not quarterly. *(Chapter 8)*

4. Our pivot triggers are written down — we agreed in advance what result would make us change course. *(Chapter 8)*

5. Our AI metrics measure business change, not adoption theater — we could distinguish a real win from a well-attended training. *(Chapter 8)*

Maturity indicator: We measure what creates results — not just the results themselves.

If you scored low: Chapter 8. Do not spend until this one is fixed.

Dimension 8: Implementation Readiness

Can we execute this successfully?

(This dimension and the next look forward — Parts Three and Four will build them. Score them now anyway; today's number is your baseline.)

1. We have selected a first initiative that is meaningful enough to matter and small enough to learn from.

2. We know what we need in an implementation partner, and how we'll evaluate one beyond the demo.

3. Change management — communication, training, inclusion of the people who do the work — is inside the implementation plan, not alongside it.

4. The people this change touches have been part of designing it, not just informed of it. *(Chapters 5, 6)*

5. We have capacity to do this — the humans involved have hours, not just enthusiasm.

Maturity indicator: Implementation is intentional — not reactive.

If you scored low: Part Three is next. Good timing.

Dimension 9: Sustainability Readiness

Will this become a capability — or a project that fades?

1. We have a plan for what happens after launch: who governs, who improves, who evaluates the next use case.

2. Saved time and early wins have a defined reinvestment path — we can say where freed capacity will go. *(Chapters 4, 8)*

3. There is a standing, fast-moving channel for employees to propose new tools and uses. *(Chapter 5)*

4. Success stories move between teams by design, not by accident.

5. We treat transformation as a capability we're building — not an initiative we're completing.

Maturity indicator: AI is becoming part of how we operate — not something we once did.

If you scored low: Part Five will meet you there.

Reading Your Results

Add up each dimension (5–25):

21-25: Strong. This dimension can carry weight. Verify with evidence, then build on it.

16-20: Developing. Workable, with named gaps. Assign owners to the specific low statements.

11-15: At Risk. This dimension will leak value from your initiative. Address it inside the initiative's first phase — scoped, budgeted, owned.

5-10: Not Ready. Stop. Spending against this dimension in this condition is how the stories in this book happened.

But the single overall number matters far less than the *shape*. Nine dimensions produce a profile — and the profile is where the honesty lives. An organization scoring Strong on Data & Technology and At Risk on Leadership isn't "pretty ready." It's an organization about to point excellent technology at an aiming system that doesn't work.

Which brings us to the most important instruction in this chapter:

Don't start with your lowest score. Start with the dimension that creates the greatest downstream impact.

Some dimensions are load-bearing. Strategic Readiness and Leadership Readiness sit upstream of everything — a wrong problem perfectly implemented is still wrong, and evidence without conditions changes nothing; you watched both happen in this book. Measurement Readiness comes next, because it's the only dimension that can tell you whether the others are actually improving. Fix upstream first. The downstream dimensions get easier when you do — which should sound familiar, because it's Chapter 7's principle applied to your own readiness.

Two final disciplines. First, *date the scorecard and keep it* — this is your organizational baseline, the "before" photo, and you'll re-score at your ninety-day mark and before every new initiative after that. The Scorecard isn't a gate you pass once. It's a compass you consult every time. Second, *watch the variance, not just the average* — where your leadership team's independent scores diverge most is where your organization doesn't agree with itself about reality. You now know what that costs.

A printable worksheet version of the full Scorecard appears in the Appendix, and a downloadable edition is available at ecfsystem.com/aiready.

Crossing Into Build

Take stock of where you now stand, because it's further than most organizations ever get.

You have a why you can say out loud. Fears named in your people's words. Shadow use mapped and read as requirements. A problem traced to its source. Two scoreboards and a dated baseline. Leaders audited against real conditions. A process tested for confusion. And now: nine scores, a profile, and a sequenced list of what to fix first.

You are no longer wondering whether you're ready. You know — dimension by dimension, with evidence.

So now we build. Part Three is where readiness becomes a plan: the right first initiative, the right partner, the roadmap, and the people-preparation that makes all of it stick. The diagnosing is done.

Turn the page. It's time to decide what to do first.

The Executive Brief

Key Insight

Impressions don't survive contact with a boardroom; scores do. The Scorecard converts ten chapters of readiness work into nine honest numbers — scored independently, backed by evidence, read as a profile. Don't start with your lowest score; start with the dimension that creates the greatest downstream impact. And keep the dated scorecard: it's the "before" photo of your transformation.

Questions Every Leader Should Ask

- If we had scored ourselves before our last major technology investment, what would we have caught?

- Where do our leadership team's independent scores diverge most — and what does that gap say about how honestly we see ourselves?
- Which of our low scores is most upstream — creating the conditions for the others?
- For every statement we rated 4 or 5: what's the evidence from the last ninety days?
- When will we re-score — and who owns making sure we do?

Monday Morning Actions

1. Distribute the Scorecard to your executive team this week. Independent scoring, no discussion first, results side by side in one working session.

2. In that session, do three things: reconcile the largest variances (ask "what evidence are you each looking at?"), identify your two most upstream weak dimensions, and assign one named owner to each.

3. Date the completed scorecard and file it with your Chapter 8 baseline. Together they are the "before" photo your entire initiative will be measured against — and the first exhibit in your ninety-day review.

Your AI Prompt

"Here are my organization's AI Ready Scorecard results: [list your nine dimension scores and your lowest-rated individual statements]. Acting as a transformation advisor: (1) identify which weak dimensions are most upstream and should be sequenced first, (2) flag any score combinations that are dangerous together, (3) for each priority dimension, suggest what evidence of improvement I should expect within 60 days, and (4) challenge me — which of my higher scores would you want proof of, and what proof would convince you?"

Reflection

Somewhere in this scorecard, you rated something a 4 that is really a 2. You know which one it is — you felt the hesitation when you scored it. What made the honest number hard to write down? That answer may be worth more than the scorecard itself.

Your 90-Day Checkpoint

Diagnose is complete when your executive team has scored independently, reconciled the variance, sequenced the gaps upstream-first, and filed the dated profile next to your baseline. You now hold something the organization in the Prologue never had: an honest, evidence-backed answer to the question this book's title asks. Part Three begins the work of acting on it.

PART THREE — BUILD

"What's the plan?"

Everything until now has been preparation and honesty: the why, the fears, the shadow map, the diagnosis, the scorecard. Necessary — and, you may have noticed, largely a study of what *not* to do. Nearly every story so far has been an autopsy.

Part Three changes the direction of the book. From here forward, we build.

Over the next four chapters you'll choose your first initiative, select the right partner, lay out your ninety-day roadmap, and prepare your people — not just your systems. The autopsies are over.

Now let me show you what the healthy version looks like.

CHAPTER 12
Choosing the Right First Initiative

The Boardroom, Rewound

I've taken you into that boardroom four times now — the receivables crisis, the collections tool, the celebration, my seat. You know every failure in it.

So this time, let's run the meeting again.

Same organization. Same billion-dollar business. Same painful receivables problem. But this time, imagine the leadership team has done the work you've just done — the why is real, the fears are named, the problem has been walked upstream, the scoreboards exist, the leaders have their conditions. The vendor suggestion still arrives, because vendor suggestions always arrive: *"There's a tool that could help us with collections."*

Here's what happens differently.

Someone asks Chapter 7's question: *if this tool works perfectly, what's still broken?* And the room answers honestly: the invoices are still inaccurate, still late, still disputed. Collections is the cleanup, not the source. The suggestion doesn't die — it gets *labeled*. A bridge, maybe, if cash demands it. Not the cure. Never the headline.

Then someone asks the question that should start every AI initiative, and it isn't about AI at all: *where does this business actually spend its people?*

The room pulls up what it now knows. In the invoicing team, capable people spend most of their week manually assembling invoices from systems that don't talk to each other — high-volume, repetitive, pattern-shaped work, exactly what Chapter 2 taught us defeats humans and suits machines. Upstream of that, contract terms flow into billing by hand. Downstream, disputes bounce between departments looking for an owner. Three candidates, all closer to the source than collections ever was.

The room does not pick the biggest one. It does not pick the flashiest one. It picks invoice accuracy — because it sits at the source, because the work is machine-shaped, because the team's own shadow users have already been experimenting with tools to fix it, because a baseline already exists from the Chapter 8 work, and because results would show up in sixty days on a number everyone already watches.

Twelve weeks later, there's a meeting with applause in it. But this time the applause is for a leading indicator: invoice accuracy up, disputes down, and — for the first time in years — the *inflow* to collections shrinking. The problem isn't being cleaned up faster.

It's being manufactured slower.

That meeting never happened. I've never had the privilege of sitting in it. But I've spent years studying exactly why it doesn't happen — and every discipline in this chapter exists to put you in that room instead of the one I sat in.

Start With the Business, Not the Candidates

Notice where the healthy version started: not with a list of AI use cases, but with a question about the business. That ordering is everything.

Most organizations choose their first initiative the way you'd expect: someone collects "AI ideas," a committee ranks them, the most exciting one wins. The list itself is the mistake — because it's usually assembled from vendor demos, conference talks, and whatever a competitor announced. It's borrowed urgency (Chapter 1) wearing a project plan.

The healthy version starts with an assessment of where your organization actually spends itself. I call it the capacity map, and it takes about two weeks of light effort:

Ask each team touching your Chapter 7 problem area to list their ten most time-consuming activities — not job descriptions, actual hours. Then sort each activity with Chapter 4's lens: is this pattern-and-volume work, judgment-

and-relationship work, or a mix? Then mark where the hours concentrate.

What you're looking for is the overlap of three things: **high hours** (enough time at stake to matter), **machine-shaped work** (pattern, volume, repetition), and **upstream position** (at or near where your problem is created, per the upstream walk). Where all three stack, you have a genuine candidate. Where your shadow discovery from Chapter 5 shows people *already* solving that work with unsanctioned tools, you have a candidate with proof of demand attached.

That's how a candidate list should be born: out of your business's actual anatomy, before a single vendor enters the room. Vendors answer questions. This is how you make sure you're asking the right one.

The Job of the First Initiative

Before the selection criteria, one reframe — because the first initiative has a different job than every initiative after it.

Your fifth AI project just has to work. Your *first* one has to work *and teach*. It is your organization's proof — not proof of concept, proof of *conditions*. Your people are watching it to learn what AI really means here: whether the AND was real (Chapter 4), whether saved time gets reinvested or harvested, whether early adopters grow or disappear, whether leadership shows up or sponsors from a distance. Whatever the first initiative demonstrates, your people will believe — and Part Four will inherit either the belief or the doubt.

That dual job is why the *right* first initiative is rarely the *biggest* opportunity on your list. Big opportunities come with cross-functional politics, long timelines, and high stakes — all of which are manageable later and lethal first. You are not choosing the initiative that transforms the company. You are choosing the initiative that earns the right to.

The Selection Filter

Run every candidate from your capacity map through seven questions. A healthy first initiative answers yes to all of them — and yes, that standard is deliberately strict, because you get exactly one first impression.

One: Is it aimed at the source — or honestly labeled? Source, bridge, or symptom (Chapter 7). Source fixes are ideal. A bridge is acceptable if you say so out loud and fund the cure alongside it. A symptom dressed as a solution disqualifies itself.

Two: Is the work machine-shaped? Task by task, is this pattern, volume, and repetition — or judgment, relationship, and accountability? (Chapters 2 and 4.) If the honest answer is "judgment," AI is the wrong tool no matter how loud the pain.

Three: Does it pass the confusion test? Can the people who do the work describe it consistently? (Chapter 10.) If not, the clarity work is phase one — scoped and budgeted — or you pick a different candidate.

Four: Can we measure it from day one? A baseline exists or can be captured now; leading indicators are defined; a pivot trigger is written (Chapter 8). An unmeasurable first initiative is a rumor with a budget.

Five: Will results show within ninety days? Not full transformation — evidence. A leading indicator that moves within a quarter. Your first initiative is also your organization's attention span; respect it.

Six: Are the people willing — ideally already ahead of you? The best first initiatives have shadow users attached (Chapter 5): people who already proved the demand, already found the value, and will become your champions instead of your resistance. Never pick a first initiative that must be imposed on a hostile team. You're proving conditions, and "imposed" proves the wrong ones.

Seven: Is it visible enough to matter and small enough to learn from? Too small — an experiment in a corner nobody sees — and success teaches nothing. Too big, and your first mistakes (there will be mistakes; progress, not perfection) happen at a scale that ends careers and appetites. You want meaningful, watchable, recoverable.

Score your candidates honestly against all seven and the decision usually makes itself. In our rewound boardroom, collections failed questions one and six. Invoice accuracy passed all seven. That's not luck — that's what selection discipline looks like when the diagnostic work has actually been done.

The Anti-Patterns

Because you'll be pitched all four of these, know them by name.

The moonshot — "let's transform the entire customer experience." Too big to learn from, too slow to show evidence, and its failure poisons the well for years. **The toy** — an AI gimmick nobody's work depends on. It can't fail, which means it can't prove anything. **The pet project** — chosen because of whose idea it was, not where the value is. Your people can smell it, and it teaches them the initiative is politics. **And the demo-driven choice** — the vendor's best demo becoming your first project, which is Chapter 7's warning in miniature: their slideware aimed at your loudest pain. The vendor should be answering your question. If you're answering theirs, stop.

If you take one sentence from this chapter, take this one: **Your first initiative is not chosen to prove the technology works. It's chosen to prove your organization can do this — honestly, measurably, and together.**

Choose it that way, and the second initiative gets easier to choose, the third easier still — because you'll have built the thing the Prologue organization never had: an organization that trusts its own aim.

Next question: who helps you build it? Because the partner you choose will either respect everything you've just done — or undo it in a quarter. That's Chapter 13.

The Executive Brief

Key Insight

The first initiative has a dual job: deliver value and teach the organization what AI really means here — it's proof of conditions, not proof of concept. Start with a capacity map of where your people's hours actually go, not a list of AI ideas. Then filter candidates strictly: source-aimed, machine-shaped, confusion-tested, measurable, fast to evidence, staffed by the willing, and sized to be recoverable.

Questions Every Leader Should Ask

- Was our candidate list born from our business's anatomy — or from demos, conferences, and competitors?
- For our leading candidate: if it works perfectly, what's still broken? Is it source, bridge, or symptom — and have we said which out loud?
- What will our first initiative teach the people watching it — about the AND, about saved time, about what happens to early adopters?
- Do we have shadow users already proving demand for this exact work?
- Is this initiative big enough to matter, small enough to survive its own mistakes?

Monday Morning Actions

1. Launch the capacity map: each team in your Chapter 7 problem area lists its ten most time-consuming activities and sorts them — pattern-and-volume versus judgment-and-relationship. Two weeks, light touch.

2. Build your candidate list only from where high hours, machine-shaped work, and upstream position overlap — then run every candidate through the seven-question filter in one working session, scored independently before discussed (you know why by now).

3. For your winning candidate, write the one-page charter: the problem at its source, the label (source or

bridge), the baseline, three leading indicators, the pivot trigger, the named owner, and the shadow users you're inviting in first.

Your AI Prompt

"Help me select my organization's first AI initiative. Here is our capacity map data: [teams, their most time-consuming activities, and which are pattern-vs-judgment work]. Here is the business problem we traced upstream: [describe]. Generate a ranked shortlist of candidate initiatives, score each against these criteria — source-aimed, machine-shaped work, process clarity, measurability, 90-day time-to-evidence, willing participants, right size to learn from — and flag which candidate is most likely to fail and why. Challenge me if my favorite looks like a moonshot, a toy, a pet project, or a vendor's demo."

Reflection

Think about the last "first project" of any major initiative in your career — the pilot, the rollout, the flagship. Was it chosen for what it could prove, or for who proposed it and how it demoed? What did the organization learn from watching it — and was that the lesson leadership intended to teach?

Your 90-Day Checkpoint

You've chosen well when you can state, on one page: where the initiative sits relative to the source, why the work is machine-shaped, its confusion-test result, the dated baseline and pivot trigger, the named owner, the shadow users invited in, and what a visible win will look like within ninety days. If any line is blank, the initiative isn't ready to fund — and finding that out today is the cheapest discovery you'll make all quarter.

CHAPTER 13
Choosing the Right Partner

The Vendor Was Never the Problem

You've already met two vendors in this book, and it's worth noticing what they had in common.

The first was in the Prologue: the collections tool. The vendor demoed against the organization's loudest pain, the tool worked, real money was collected — and nothing that created the problem changed. The second was in Chapter 2: the reconciliation platforms. Serious vendors, real AI capability, sitting in demo environments for two years while leadership circled a question it didn't understand.

Two failures. Zero of them the vendor's fault.

The collections vendor answered the question the organization asked — the organization asked the wrong one. The reconciliation vendors waited patiently while a leadership team vetted without knowing what a good answer would look like. In both rooms, the vendor was a mirror. It reflected back exactly the level of preparation the organization brought to the table.

That's the truth underneath every partner decision, and it's why this chapter sits *after* eleven chapters of internal work instead of at the front of the book, where most organizations start:

An unprepared organization gets sold. A prepared organization selects.

Vendors are not villains — the good ones genuinely want your problem solved, because solved problems become references. But every vendor, honest or not, can only aim at the target you show them. Show them your pain, and they'll sell to your pain (Chapter 7 told you the whole market is organized that way). Show them your *source* — the problem, the process, the baseline, the people — and something different happens: the conversation stops being a pitch and becomes an evaluation. Theirs of the problem. Yours of them.

So before we talk about choosing a partner, we have to talk about what you bring into the room.

Before the First Demo: Go Inside First

Here is the step organizations skip, and it's the one I want to slow down on because everything else in this chapter depends on it.

Before you evaluate a single partner, go deep *inside* — internal conversations, real deep dives, with the people on the front line of the problem you chose in Chapter 12. Not their managers. Not a summary of their tickets. Them.

The people on the front line have a view of your business that you will never have from your seat — and I mean *never*. They know where the process actually breaks, which workaround is holding the quarter together, which customer complaint is really a process defect wearing a customer's voice. Yes, there will be noise in those conversations. There will be complaints, some of them old, some of them personal. Sit through it, because inside the noise is something most leaders never stay long enough to hear: the solution. These people aren't just speaking for themselves.

They're speaking for your customers.

Every frustration a frontline employee carries is a frustration your customer felt first or will feel next — the late invoice, the wrong answer, the fourth transfer, the "let me check on that." When you deep-dive with the front line, you're not doing employee engagement. You're doing customer research and process discovery at the same time, from the only people who see both.

You've built the tools for this already. The four-question meeting from Chapter 3. The shadow discovery from Chapter 5. The confusion test from Chapter 10. The capacity map from Chapter 12. This is where all of it gets spent — because the output of those conversations is the thing you carry into every partner meeting:

The right problem.

Say it plainly: **before you choose the right partner, choose the right problem.** Understand the processes. Understand the people. Understand — and this matters just as much — the *lack* of processes, the places where the work runs on tribal knowledge and duct tape, because that absence is exactly what a partner needs to know and exactly what an unprepared buyer hides without realizing it.

Walk into the vendor conversation with your one-page charter from Chapter 12 — the problem at its source, the label, the baseline, the indicators, the pivot trigger — and you have changed the physics of the meeting. They can't sell you your symptoms. You already know better.

Partner Versus Collaborator

Now, language — because the word "partner" gets printed on every vendor's slide, and it's earned by almost none of them.

A **vendor** sells you a tool. The relationship peaks at the contract signature.

A **partner** delivers a project. The relationship peaks at go-live.

A **collaborator** joins your problem. The relationship peaks when the *business number moves* — because that's the outcome they signed up for.

What you want, for a first initiative especially, is the third one — and the difference shows up early if you know what to look for. A collaborator asks more questions than they answer in the first meeting. A collaborator wants to talk to the people who do the work, not just the people who sign the check. A collaborator gets visibly more interested when you show them your confusion-test results — because real ones know that's where implementations die, and they're relieved to meet a buyer who knows it too. And a collaborator will tell you a truth that costs them: *this part isn't ready,* or *that expectation is wrong,* or even *we're not the right fit for this piece.*

A vendor tells you what you want to hear. A collaborator tells you what you need to know. You can hear the difference in a single meeting — if you ask the right questions.

The Questions That Reveal a Collaborator

So ask these — and listen as much to *how* they're answered as to what's said.

"Tell me about solutions similar to ours that you've solved." Not their flagship case study — *your shape of problem*. Same kind of source, same kind of process mess, comparable scale. Then ask what happened after go-live: did the business number move, and can you talk to that client? A collaborator has specifics and offers the reference before you ask twice. A vendor has a highlight reel from a different industry and a story that always ends at deployment.

"What are we not seeing?" This is the honesty test, and it's my favorite question in the set. You've done eleven chapters of homework — invite them to find the hole in it. A collaborator will find one, because there's always one, and the quality of what they find tells you the quality of who you're hiring. A vendor answers, "Honestly, you're further along than most clients we see." That's not a compliment. That's a sales motion.

"What does success look like to you?" Listen very carefully here, because their honest answer is a preview of the next twelve months. If success is *deployment, adoption numbers, licenses activated* — that's Chapter 8's adoption theater, offered as a deliverable. What you want to hear is your own scoreboard coming back at you: the leading indicators, the source metrics, the business outcome. Better yet: a partner who asks to see *your* definition of success before offering theirs.

Then add these four:

"What would make you tell us not to buy?" A collaborator has an answer, because they've walked away from bad-fit deals and will tell you why. A vendor has never met a problem their platform doesn't solve.

"What will you need from us to succeed?" This one tests whether they understand conditions. The right answer includes your time, your data, your decision speed, access to your frontline people, and an executive who shows up — Chapter 9's list, coming from their side of the table. If they need nothing from you, they're planning to install software, not change outcomes.

"Who, exactly, will do the work?" The people in the sales meeting are rarely the people on the project. Meet the actual team before you sign.

"Walk me through an implementation that went badly — and what you changed because of it." Everyone has one. Only collaborators will tell you about it. The ones who've never failed are either brand new or lying, and neither belongs on your first initiative.

What You Owe Them

One more thing, because partnership runs both directions and this book doesn't do one-sided accountability.

If you want a collaborator, you have to be one. That means honesty about your conditions — show them the confusion-test results, the scorecard gaps, the change debt from Chapter 6, the real state of your data. It means access — to your frontline people, your process reality, your decision-makers. It means decision speed — a partner who waits four weeks for every answer is being taught that your organization isn't serious (you learned in Chapter 5 what your own people conclude from slow governance; partners conclude the same). And it means an executive sponsor who actually shows up, because partners calibrate their effort to your visible commitment just like employees do.

Hiding your mess from a partner is like hiding symptoms from a doctor. The engagement will be pleasant, the diagnosis will be wrong, and you'll pay for both.

Choose the right problem. Bring it honestly. Ask the questions that only collaborators can pass. Do that, and the partner decision — the one that undoes so many organizations — becomes one of the easier calls you'll make this year.

Now you have the initiative and the collaborator. What you need next is the plan — the ninety days themselves. That's Chapter 14.

The Executive Brief

Key Insight

The vendor is a mirror: an unprepared organization gets sold; a prepared organization selects. Before you choose the right partner, choose the right problem — through frontline deep dives, because your people's noise contains your customers' voice. Then hire a collaborator, not a vendor: one who joins your problem, asks for your conditions, and defines success by your scoreboard, not their go-live.

Questions Every Leader Should Ask

- Have we gone deep with the frontline people who live this problem — and did we stay long enough to hear the solution inside the noise?
- Are we walking into vendor meetings with our charter, baseline, and confusion-test results — or with our loudest pain?
- Does this partner's definition of success match our scoreboard, or their deployment plan?
- What did they say when we asked what we're not seeing — and did the answer cost them anything?
- Are we prepared to be a collaborator ourselves: honest about our mess, fast with decisions, present at the working sessions?

Monday Morning Actions

1. Before any demo, schedule the frontline deep dives on your chosen initiative — the people who do the work, ninety minutes, using the Chapter 3 questions plus one more: "If you could fix one thing about how this works, what would it be?" Write down everything, including the complaints. Especially the complaints.

2. Build your partner evaluation packet: the one-page charter, the baseline, the confusion-test summary, and your definition of success. This is what every candidate partner sees — and their reaction to it is your first data point.

3. Run every serious candidate through the seven questions in this chapter, with the same people asking them each time. Score independently before comparing (you know why by now). Check the references whose problems looked like yours — and ask those references what the partner needed from them.

Your AI Prompt

"I'm evaluating implementation partners for an AI initiative. Here's our one-page charter: [paste]. Help me prepare: (1) generate follow-up questions specific to our problem for each of these areas — similar solutions delivered, what we're not seeing, their definition of success, when they'd advise not buying, what they need from us; (2) list the red-flag answers I should listen for; and (3) tell me what a genuinely strong answer would sound like for each, so I can tell a collaborator from a vendor in one meeting."

Reflection

Think about the best working relationship you've ever had with an outside firm — and the worst. The difference probably wasn't their talent. What did the best one know about your business that the worst one never asked? Who let that happen — them, or you?

Your 90-Day Checkpoint

You've chosen well when your partner can state your problem in your words, your success in your metrics, and your gaps without your prompting — and when your frontline people say "they actually listened to us" after the discovery sessions. If your partner's project plan could be copy-pasted onto another company, you haven't hired a collaborator. You've bought a deployment.

CHAPTER 14
Your 90-Day Roadmap

Two Years Becomes Ninety Days

In Chapter 2, I told you about an organization that talked about the same problem for two years. Well-defined problem, capable people, vendors waiting — and meeting after meeting that felt like diligence and functioned like avoidance. I made you a promise in that chapter: with the right understanding, *two years becomes ninety days.*

This chapter is where I keep that promise.

Everything you've built across this book — the why, the fear conversation, the shadow map, the upstream walk, the scoreboards, the leadership audit, the confusion test, the Scorecard, the initiative charter, the partner questions — is about to become something none of it has been yet: a calendar. Because there is a moment in every transformation I've ever led where the thinking has to take a date, or it quietly becomes the two-year conversation. Plans without dates are opinions.

First, let's be honest about what ninety days is and isn't.

Ninety days will not transform your organization. Anyone who promises that is selling you something — and now you know the questions to ask them. What ninety days *will* do is complete one full, honest cycle: get truthful, get ready, get moving, and get evidence. One initiative, taken from diagnosis to measurable signal. That cycle is the thing your organization has never had — and once it exists, it repeats. The transformation isn't the ninety days. The transformation is becoming an organization that can *run* ninety-day cycles.

Why ninety? Because it matches how organizations actually hold attention. Long enough for a leading indicator to move (Chapter 8 said sixty days; we're giving it room). Short enough that sponsors stay engaged, skeptics stay curious, and mistakes stay cheap — the compounding-learning argument from Chapter 1, made operational. And short enough that nobody can hide. In a two-year plan, month three looks like every other month. In a ninety-day plan, day thirty is a gate with your name on it.

Three rules before the calendar:

The sequence is fixed; the dates are yours. A smaller organization might run this in sixty days; a complex enterprise might need a hundred and twenty. Stretch the timeline if you must — never reorder it. Organizations that skip a phase don't move faster. They create risk (you've now read eleven chapters of the receipts).

One initiative. Not a portfolio. Not "three quick pilots." One — chosen by Chapter 12's filter. The first cycle is proof of conditions, and conditions are proven one honest initiative at a time.

Every phase has an exit gate. You don't advance on the calendar. You advance on the evidence. If day thirty arrives and the gate isn't met, the calendar moves — not the standard.

Days 1-30: Get Honest

The first month is Awareness and Diagnose, run as an operation instead of a book you happened to read. It's the busiest month for the executive team and the cheapest month of the entire journey — nothing is purchased, everything is discovered.

Week one — the why and the fear. Hold the two-hour, no-technology-on-the-agenda session from Chapter 1: constraints, capacity, what AI should never touch, and a why every executive can say out loud. Then hold the fear meeting from Chapter 3 with the team closest to your likely rollout — four questions, no slides, and answers you write down in their words. Start the communication drumbeat (Chapter 6) *this week*, before there's anything to announce: "Here's what we're exploring, here's why, here's how you'll hear about it." The rumor mill starts on day one whether you do or not.

Week two — the discovery. Declare the Shadow AI amnesty and run the discovery from Chapter 5: what's in use, for what, and what it's worth. In parallel, launch the capacity map from Chapter 12 across the teams touching your biggest pain: ten most time-consuming activities each, sorted pattern-versus-judgment.

Week three — the walk and the numbers. Run the upstream walk (Chapter 7) on your leading problem with the people who touch the work. Capture the baseline at the source and draft both scoreboards with pivot triggers (Chapter 8). Run the confusion test (Chapter 10) on the process the walk points to.

Week four — the mirror. Executive team scores the AI Ready Scorecard™ independently; one working session to reconcile variance, sequence the gaps upstream-first, and assign owners (Chapter 11). Audit the leadership conditions — sponsor named, transformation leader's five conditions scored, adoption and communication owners written down (Chapter 9).

The Day-30 gate. You may proceed when: the why is one consistent sentence; the top three fears are named in your people's words; the shadow map exists; your problem is traced to its source; a dated baseline and both scoreboards exist; the Scorecard profile is filed; and the sponsor, transformation leader, and owners are named — with conditions, not just titles. Miss any of these and the honest move is to extend Diagnose, not to shop anyway. The most expensive thing you can do in the next sixty days is solve the wrong problem efficiently.

Days 31-60: Get Ready

The second month is Build. This is where most organizations start — which is why most organizations fail. You're starting here with a month of truth already banked.

Weeks five and six — choose and charter. Build the candidate list from the capacity map overlap (high hours, machine-shaped, upstream). Run the seven-question selection filter (Chapter 12), scored independently before discussed. Write the one-page charter for the winner: problem at its source, honest label, baseline, three leading indicators, pivot trigger, named owner, and the shadow users you're inviting in first. If the confusion test flagged clarity work, it gets scoped, budgeted, and owned here — phase one of the initiative, not a reason to stop.

Weeks six through eight — select the collaborator. Frontline deep dives first (Chapter 13), then the evaluation packet goes to candidate partners, then the seven partner questions — same askers each time, scored independently, references checked against your shape of problem. You are not looking for the best demo. You're looking for the firm that gets *more* interested when you show them your confusion-test results.

Weeks seven and eight — prepare the people. This runs in parallel, and the next chapter is devoted to it, so here I'll say only what belongs on the calendar: the initiative's why communicated before its what (Chapter 6, applied); the AND made concrete for the affected team — what the machine will carry, what they keep, where freed time goes (Chapter 4, applied); access, time, and safety funded, not implied; and your shadow users converted into named champions with real roles.

The Day-60 gate. You may proceed when: the charter is signed; the partner is chosen and can state your problem in your words and your success in your metrics; the affected team has heard the why from leadership directly — before the tool arrives; the champions are named; and the measurement plan has an owner and a standing meeting. If you have a partner but unprepared people, you are not ready — you're just accompanied.

Days 61-90: Get Moving

The third month is Execute — and the shortest section of this chapter, because Part Four is entirely about it. What belongs here is the rhythm.

Launch small and visible — with the willing team, at the scoped process, exactly as chartered. Leaders present in the working sessions, using the tools themselves, because your people watch feet, not memos.

Review rhythm weekly. The leading indicators from your scoreboard, reviewed at the speed of the work — not the quarter. Is it safe to experiment? Are use cases spreading? Where did saved time go? What's the age of the oldest unresolved blocker? That last one gets a standing agenda line: barriers surfaced this week, barriers removed this week, and the oldest one still breathing — with a name attached to killing it.

Feed the drumbeat. What's working, what we're hearing, what we're fixing, what changed because someone spoke up. Wins travel by design: the person who found the improvement tells the story, publicly, and grows for it — because the whole organization is watching what happens to the first adopters.

The Day-90 gate — the honest review. Baseline versus today, at the source. Leading indicators against the sixty-day expectations you wrote in advance. Pivot triggers checked — *before* the celebration, not instead of it. And then one of three decisions, all of them wins: **scale it** (evidence says expand — carefully, Part Five will show you how), **adjust it** (signal is real but aimed slightly wrong — rechart and run again), or **pivot** (the pre-agreed trigger fired — you stop, say so out loud, and spend the learning on cycle two). The only losing outcome is the fourth one, the one you've now been inoculated against: ambiguity with applause.

Whatever the decision, one more act closes the cycle: go back to the people who raised concerns in week one and show them the evidence — moved, resolved, or still open. The acknowledgment loop from Chapter 9, kept. That single habit, you'll recall, would have changed the ending of every story in this book.

Day 91

Here's what you actually built, and it's more than one initiative's result.

You built an organization that has now done the loop once: gotten honest, chosen well, measured truthfully, and faced its own evidence on a schedule. The second cycle starts with assets the first one didn't have — a dated Scorecard to re-score against, a proven champion bench, a partner who knows your business, a drumbeat people now trust, and an executive team that has practiced saying "this isn't working" without anyone dying.

That's why the title of this book was never really about the ninety days. Ninety days is the training cycle. The capability is the prize — and making it permanent is what Part Five is for.

But first: none of this survives contact with reality if your people aren't ready for it. Systems get installed. People get prepared. That's Chapter 15.

The Executive Brief

Key Insight

Ninety days doesn't transform the organization — it completes one honest cycle: get truthful (days 1-30), get ready (31-60), get moving (61-90), and face the evidence at three hard gates. The sequence is fixed; the dates are yours. Plans without dates are opinions, and organizations that skip phases don't move faster — they create risk.

Questions Every Leader Should Ask

- Does our current AI plan have dates and exit gates — or is it a two-year conversation wearing a project name?
- What would have to be true by day 30 for us to honestly proceed — and are we willing to hold the calendar hostage to the evidence?
- Are we running one chartered initiative, or scattering ourselves across a portfolio of unproven pilots?

- Who reviews the weekly rhythm in days 61–90, and what's the age of our oldest unresolved blocker?
- At day 90, are we prepared to scale, adjust, or pivot — and have we pre-agreed which evidence forces which?

Monday Morning Actions

1. Put the three gates on the executive calendar today — day 30, day 60, day 90 — with attendance mandatory for the sponsor. The gates are the roadmap; everything else is scheduling.

2. Assign each week-one activity an owner and a date: the why session, the fear meeting, the amnesty announcement, and the first drumbeat communication. Four events, week one, no exceptions.

3. Write the day-90 review agenda *now*, before launch: baseline versus actual at the source, leading indicators versus expectations, pivot triggers checked, and the scale/adjust/pivot decision. Publishing the exam before the course is what keeps everyone honest during it.

Your AI Prompt

"Build me a detailed 90-day organizational AI readiness plan from this framework: Days 1–30 (why session, fear conversation, shadow discovery, upstream walk, baseline and scoreboards, leadership audit, confusion test, readiness scorecard), Days 31–60 (initiative selection and charter, partner evaluation, people preparation), Days 61–90 (scoped launch, weekly leading-indicator rhythm, barrier removal, day-90 review). My context: [organization size, industry, the problem we've chosen, key constraints]. Produce a week-by-week calendar with owners, meeting cadences, the three gate checklists, and the two biggest risks to this timeline given my context."

Reflection

Look back at the last major initiative your organization launched. If it had faced a hard gate at day 30 — evidence required, calendar hostage to the standard — would it have passed? Would it have ever launched at all? What does that answer tell you about what your gates have been protecting: the initiatives, or the people who proposed them?

Your 90-Day Checkpoint

This chapter *is* the checkpoint — so here it is in one line: three dated gates on the executive calendar, week one's four events owned and scheduled, and the day-90 exam published before the course begins. If those exist, your ninety days have started. If they don't, you're still in year one of the two-year conversation.

CHAPTER 15
Prepare People, Not Just Systems

Three to One

I need to take you back to the integration one more time — the acquisitions, the vocabularies, the cutovers. You know that engagement from Chapters 6 and 10. But there's one observation from it I've saved for here, because it belongs at the end of Build, right before you launch anything.

When I was brought in, I started doing what I always do: mapping where the organization's attention was going. And one ratio struck me almost immediately.

The time being spent preparing the *system* was getting roughly three times the attention of the time being spent preparing the *people*.

Three to one. System workstreams had plans, owners, meetings, milestones, vendor support, and a cutover date circled on the calendar. The people side had... a training session. Somewhere near the end. If there was time.

And I remember knowing, in that moment, that we were going to be in trouble. Not because the system work was wrong — it was necessary and mostly well done. Because of a truth I had learned across every transformation I'd ever touched:

The system does what we tell it to do. The people do what they have to do.

Read that again, because the entire chapter lives inside it. A system is obedient. Configure it, and it behaves — every time, forever, exactly as instructed. People are not obedient, and thank goodness, because obedience isn't what you hired them for. People are *adaptive*. Put a person in front of a customer with a tool they don't understand, a vocabulary nobody taught them, and a job that still has to get done in the next four minutes — and they will do what they have to do. They'll reach for the old habit. They'll invent a workaround.

They'll guess at the field (and you know from Chapter 6 what every guess becomes). They'll keep the customer served and the old world alive, inside your brand-new system.

Here's how that story ended — or rather, how it didn't. A year later, the organization was still talking about fragmented data. Not all of the acquired companies had been brought into the new system, the new vocabulary, or the new way of thinking about how the systems ran and how customers were served. And to this day, I would honestly guess that the full integration has never been finished.

Sit with that. The system went live. The integration didn't finish. Those are two different events, and most organizations only plan for the first one.

Go-live is a system event. The change is a people event. You can complete a system and never complete the change — and then spend years wondering why the "finished" project keeps generating fragmented data, workarounds, and meetings about problems you thought you solved.

What Was Actually Happening

Why does the ratio tilt three to one? Not because leaders don't care about people. Because of how the two kinds of preparation *behave* in a project plan.

System preparation is legible. It has tasks a vendor can quote, milestones a PMO can track, and a testing phase that fails *loudly, before launch* — a broken integration announces itself in UAT, gets a ticket, gets fixed. Everything about system work rewards attention, so attention flows to it.

People preparation is illegible. No vendor sells it by the module. Its milestones feel soft ("understanding"? "fluency"? put *that* on a Gantt chart). And critically, it fails *quietly, after launch* — an unprepared workforce doesn't crash on cutover day. It smiles, attends the training, and then does what it has to do. The failure shows up months later, dressed as something else: data quality issues, adoption problems, "resistance," a resurgent Shadow AI. By then, nobody connects it to the preparation budget that got cut a year earlier.

So the three-to-one ratio isn't a values problem. It's a visibility problem — attention flows to the work that shows up in dashboards, and away from the work that only shows up in outcomes. Which means fixing it isn't about caring more. It's about making people-readiness as legible, planned, and gated as system-readiness. That's this chapter.

One more thing before the how. If you think AI makes this less important, it's the opposite. Every generation of technology gets *easier to deploy* — and AI is the easiest yet. The system side of an AI initiative can be ready in weeks now. Which means the long implementation runway that used to give people months to hear about, absorb, and adjust to a change has collapsed. The technology's speed has quietly made people-preparation the critical path of every AI initiative. The easier the tools get, the more this chapter matters.

What Prepared Actually Means

Now let's define the work, because here's the trap: most organizations hear "prepare the people" and book a training class. Training is real, and it's the *smallest* part. A class transfers information. Readiness is built somewhere else. Prepared people have five things, and by now you'll recognize where each one comes from — this section is Part One and Part Three, converging on launch day.

Understanding. They know the why — before the what, from leadership's mouth, more than once (Chapter 6's drumbeat, aimed at this initiative). They know what's changing, what it means *for them*, and what the new words mean. Nobody logs in one morning to find their vocabulary gone. That is the direct, personal repayment of the story that opened Chapter 6.

Fluency. Not exposure — fluency. That requires Chapter 4's three payments in full: access to the actual tools before the stakes are real, time carved out to practice (not "learn it on top of everything else"), and safety to be publicly clumsy. A rehearsal, not a lecture: real scenarios, real records, a day-in-the-life run-through before the day in the life arrives.

Fingerprints. The people doing the work helped design the change — the frontline deep dives from Chapter 13, the shadow users from Chapter 5 converted into named champions, the workaround inventory from Chapter 10 actually addressed rather than automated over. People don't resist what they helped build. They defend it.

Support at the moment of need. The most underrated line item in any launch: floor support. Not a helpdesk number — humans, present, in the first weeks, at the moment someone is staring at a screen with a customer waiting. Because that moment is exactly where "people do what they have to do" gets decided. If help is thirty seconds away, they do the new thing. If help is a ticket queue away, they do the old thing — and the old thing compounds.

A new definition of good. Their measures, expectations, and even job descriptions updated to match the new work. If the machine now does the matching and the human handles exceptions and judgment (Chapter 4's AND), but the person is still measured on the old volume metric, you have ordered them to fail one scoreboard or the other. People do what they have to do — and what they have to do is whatever you *measure*. Align it before launch, or it will misalign them after.

And follow that thread one step further, because it runs deeper than metrics: *the roles themselves are changing* — and most organizations leave the paperwork of the workforce frozen in the old world. Job descriptions still describe the work the machine now carries. Hiring profiles still screen for yesterday's skills. Interview questions still test for the volume work instead of the judgment work. New-employee training still teaches the process that no longer exists. Every one of those documents quietly hires, evaluates, and onboards people for a job you've already redesigned — which means you'll keep acquiring the old workforce while asking it to do the new work.

This is not something the transformation leader can fix alone, and it's why one partnership belongs explicitly in your plan: **HR working hand in hand with the transformation executive** to clarify what each affected role has actually become — what the machine carries, what the human keeps, what new skills the pairing requires. And from there, transformation works with leadership at *every level* — because a role's reality lives with the leader who manages it, not in a template — to make sure those changes land in the real artifacts: the job descriptions, the hiring criteria, the interview guides, the onboarding curriculum, the career paths. When HR is missing from the transformation table, the org chart keeps regenerating the old organization underneath the new initiative, one hire and one review cycle at a time.

Making It Legible: The Parity Rule

Here's how you institutionalize all of this, and it's simple enough to fit on a sticky note:

Every system milestone gets a people twin.

Data migrated? Twin: vocabulary taught and tested — pull ten records, ask three people what the fields mean, score the agreement (you know this test). System configured? Twin: champions fluent and certified to support their teams. UAT passed? Twin: day-in-the-life rehearsal passed — real people, real scenarios, observed. Cutover scheduled? Twin: floor support staffed and the week-one drumbeat written. Hypercare running? Twin: feedback loop live, with visible fixes — "here's what you told us, here's what changed."

Then audit your plan the way I audited that integration: count the meetings, the owners, the calendar days, and the dollars on each side of the ledger. It doesn't need to be one-to-one in cost — systems are genuinely expensive. It needs to be one-to-one in *seriousness*: named owners, dated milestones, and gate criteria on the people side, just like the system side. If your day-60 gate (Chapter 14) can be passed with an unready workforce, your gate is broken.

The test I'd leave with every executive is one question, asked in the launch-readiness review: *"Show me the people plan at the same altitude as the system plan."* If one is a bound document and the other is a training date, you already know your ratio. And now you know what the ratio costs — because somewhere out there is an integration, years old, still not done, still generating fragmented data, because a system got three times the attention of the humans who had to live in it.

Closing Part Three

Look at what Build has given you.

An initiative chosen from your business's actual anatomy, filtered strictly, and chartered on one page (Chapter 12). A collaborator who can state your problem in your words and your success in your metrics (Chapter 13). A ninety-day calendar with three hard gates (Chapter 14). And now a workforce that is prepared, not just notified — understanding, fluent, fingerprinted, supported, and measured for the new work.

The plan is built. What remains is the part that no plan can do for you: leading it, week by week, while your entire organization watches what you actually do.

That's Execute. Turn the page.

The Executive Brief

Key Insight

The system does what we tell it to do; the people do what they have to do. System readiness fails loudly before launch; people readiness fails quietly after it — so attention flows three-to-one toward the system, and the change never actually finishes. Go-live is a system event; the change is a people event. Give every system milestone a people twin, gated with the same seriousness.

Questions Every Leader Should Ask

- In our current plan, what ratio of meetings, owners, and dollars goes to system preparation versus people preparation? Would the answer embarrass us?
- Could our launch gate be passed with a technically perfect system and an unready workforce?
- Do the people affected have understanding, fluency, fingerprints, moment-of-need support, and updated measures — or a training date?
- If someone is stuck with a customer waiting in week one, how many seconds away is help — and what will they do while they wait?
- Are we still measuring anyone on the old definition of good while asking them to work the new way?
- Is HR at our transformation table — and do our job descriptions, hiring criteria, and new-employee training describe the roles as they now exist, or as they existed before the machine arrived?

Monday Morning Actions

1. Audit the ratio: take your initiative's project plan and count both sides — system tasks versus people tasks, with owners and dates. Publish the number to the executive team. The ratio itself will make the argument.

2. Build the twin list: for every system milestone on the plan, write its people twin, assign it an owner, and add it to the same gate. One page, this week, before the day-60 gate.

3. Fund floor support now — named humans, physically or virtually present, for the first three weeks after launch. It's the cheapest insurance in this book: the cost of a few people's time against the compounding cost of a workforce quietly deciding the old way is safer.

Your AI Prompt

"Here is the project plan for our upcoming AI initiative: [paste or summarize milestones]. Analyze the balance

between system preparation and people preparation. Then: (1) for every system milestone, propose a 'people twin' milestone with clear pass criteria; (2) identify the moments of need — points in the first 30 days where a user under pressure is most likely to revert to old methods — and design support for each; (3) flag any role whose current metrics or incentives conflict with the new way of working. Assume people do what they have to do, not what they're told to do."

Reflection

Think of a "completed" project in your organization's past that never quite finished — the system that went live years ago but still generates workarounds, complaints, and meetings. Nobody calls it a failure; nobody calls it done. What ratio of attention did its people side get? And who is still, quietly, doing what they have to do to keep it running?

Your 90-Day Checkpoint

Part Three is complete when four things exist on paper: the one-page charter, the signed collaborator who passed the seven questions, the dated roadmap with three gates — and a people plan at the same altitude as the system plan, with twins, owners, and floor support funded. If the fourth one is missing, you haven't finished Build. You've finished procurement.

PART FOUR — EXECUTE

"Are we leading it well?"

The plan is built. The initiative is chartered, the collaborator chosen, the roadmap gated, the people prepared. What happens next is the part no document can do for you.

Execution is where your organization stops listening to what you say and starts watching what you do. Every belief you asked your people to hold in Part One — the why, the AND, the safety — now gets tested against your actual behavior, week by week, in rooms you're not in.

The next four chapters are about passing that test: earning adoption instead of announcing it, leading visibly instead of sponsoring from a distance, moving your people from using AI like a tutor to working with it like a teammate, and running the weekly rhythm that keeps momentum alive and barriers dead.

The plan was the easy part. Now we execute.

CHAPTER 16

Adoption Is Earned, Not Announced

Seven Years of Promises

One of the most instructive chapters of my career came at a software company — and it started with a promise the company had been making for over seven years before I ever walked in the door.

The promise was a learning platform. Something that would help customers learn the product, elevate adoption, and make onboarding smoother. Year after year, iteration after iteration, the organization had tried to get it built — with minimal understanding of what it actually takes to build a learning program. Each attempt stalled. The promise stayed on the roadmap. The customers kept waiting.

Here's the assumption underneath all those stalled attempts, and I want you to check whether it lives in your organization too: *if you create something for training, people will take it, adhere to it, and adopt it.*

They won't. Whether people realize it or not, there is a science behind teaching adults, just as there is a science behind teaching children — it's why the field has two names, pedagogy for children and andragogy for adults. Adults don't learn because content exists. They learn when the material connects to their actual work, respects the experience they already carry, solves a problem they have *this week*, and arrives in a way that lets them practice, stumble, and ask questions safely. Build a library and announce it, and what you've built is content. Capability is a different construction project entirely.

I came into that organization carrying a deep background in learning and development — not just an executive title — and I understood and respected that science. So when I was asked to finally get the platform across the finish line, I was genuinely excited. I knew I was fully capable. And I knew exactly what I would need: not budget, not tools — *people*. Access to the individuals who held the knowledge, and a working partnership with the teams closest to the customer, so the platform would be built alongside them instead of delivered to them.

Here's what I got instead: one individual assigned to help.

And here's the decision that still stands out to me as one of the worst I've watched an organization make: the people who worked directly with customers during onboarding — one of the largest groups in the company, the very group whose daily work this platform would touch most — were excluded from the first year of the rollout.

Sit with the irony. Those were precisely the people who could have elevated adoption — they had the customer relationships, the credibility, and the practical knowledge of where learning actually broke down. Properly integrated, the platform could even have *shortened* client onboarding, which was their daily grind. The tool was built to make their world better. Nobody invited them into building it.

Nobody asked what impact a tool like this would have on the organization — the fear it might surface, the concern, the frustration, the confusion. A new tool is never neutral to the people whose work it touches; it arrives carrying an implied judgment about how they've been doing things. Unasked and unincluded, that group met the platform with exactly what you'd now predict: really low adoption and really high frustration — in the group responsible for implementation.

And the organization had never measured adoption to any real degree, so there wasn't even a scoreboard to catch it happening.

The ending writes itself, and it's the sentence I want you to remember longest: because adoption was low *internally*, adoption and utilization were even lower for our *customers*.

We built a platform to teach adoption. And we failed to practice it.

And now I owe you a confession from that engagement, because the biggest lesson in it was mine.

I assumed the buy-in. The promise was seven years old. The organization had remarkably low turnover. I put those two facts together and read them as consent: surely, after all this time, with all these long-tenured people, everyone wanted this thing. So I didn't go looking for resistance at the start.

I discovered it after the tool was built — and not from the inside. From *customers*. In conversations with them, I learned the platform wasn't even being mentioned during implementation and onboarding. The group that had been excluded from the build had quietly returned the favor: in their conversations with customers, the tool simply didn't exist. I had to work backwards from there — gathering the feedback I should have had on day one, retrofitting inclusion onto a finished product.

We eventually got there. But I won't pretend the detour was free. It cost roughly eighteen months — and it cost more than time. If I'd had that frontline feedback at the beginning, the tool itself would have been more robust. Adoption would have been higher. We'd have had clear KPIs telling us exactly what to do next, instead of discovering the gaps by archaeology.

So let me hand you the vow I made walking out of that experience: never again will I assume — based on the age of a promise, the tenure of a workforce, or the silence in the room — that adoption has been discussed, let alone taken care of. Assumed buy-in is the most expensive kind, because you don't find out you never had it until the moment you need it.

What Was Actually Happening

Strip the story to its mechanics and you find the belief this chapter exists to kill.

The organization treated adoption as an *event* — something that happens at launch, activated by an announcement. Build the thing, ship the thing, tell people about the thing, done. Seven years of iterations, and every one of them carried that same assumption in its foundation.

But adoption is not an event, and it cannot be announced into existence. **Adoption is a verdict.** It's the judgment your people deliver — individually, privately, in the flow of their actual work — on everything you did before and after the launch. Did you include them or surprise them? Does this solve a problem they have, or a problem someone imagined they have? Is it easier than their workaround? Is it safe to be clumsy with it? Did anyone ask? Every person weighs those questions, mostly unconsciously, and renders a verdict with their behavior. Announcements don't sway that verdict. Evidence does.

Look at the specific evidence my story's organization presented to its own people:

You weren't worth including. The largest affected group, locked out of year one — which taught them (Chapter 6) that decisions about their work happen without them. *Your knowledge isn't needed.* One person assigned to a build that needed dozens of contributors — so the platform couldn't reflect the reality of the work, and everyone close to the work could tell. *Your feelings about this are irrelevant.* No one asked what the tool would stir up — so fear and frustration had nowhere to go but into resistance (Chapter 3, verbatim). *And nobody's checking anyway.* No adoption measurement — so quiet non-use carried no signal and no consequence until the customer numbers arrived, far too late (Chapter 8, in miniature).

Given that evidence, low adoption wasn't a failure of the workforce. It was a *correct reading of the situation*. Your people's adoption behavior is almost always rational — it just responds to what you did, not what you announced.

Now let's dissect my assumption, because you will be tempted by the same logic, and it's worth knowing exactly where it breaks. A seven-year-old promise doesn't create appetite. It creates skepticism — the workforce had watched iteration after iteration stall, and every stall deepened the change debt (Chapter 6). And low turnover doesn't mean buy-in. It means *long memories* — the people still in their seats were the very people who had lived through every previous attempt. Tenure doesn't measure enthusiasm for your next initiative. It measures how much history your next initiative has to overcome.

And notice where the truth finally surfaced: outside the building. When internal resistance is so quiet that you first hear about it from your customers, you don't have a resistance problem — you have a channel problem. The silence you're reading as consent is just resistance with nowhere safe to go. Chapter 3 warned you about that silence. I'm the proof that even people who know better can mistake it for agreement.

And then there's the cascade, which deserves its own paragraph because almost nobody plans for it: **your customers will never adopt what your people haven't.** Every customer-facing tool passes through your employees' hands, voices, and enthusiasm on its way to the customer. If your onboarding team is frustrated and half-fluent, that is precisely the experience your customers receive — at a discount. Internal adoption isn't a nice-to-have on the way to customer adoption. It's the ceiling on it.

How Adoption Is Actually Earned

So how do you earn the verdict? You already hold most of the tools — this is where they go to work. Five earning mechanisms, in the order they matter:

Earned through fingerprints. People defend what they helped build (Chapter 15). The group my story's organization excluded is the group you invite in first — not to a feedback session after decisions are made, but to the design table. Your shadow users and champions (Chapters 5 and 12) are the advance guard; the broader affected team follows through structured involvement: reviewing workflows, testing early versions, naming what's wrong while it's still cheap to fix.

Earned through relevance. This is the andragogy, made practical. Adults adopt what solves *their* problem *in their workflow*. So enablement is problem-first, not feature-first: start from the ten most time-consuming activities on their capacity map (Chapter 12) and show, concretely, which ones get lighter. Teach in the flow of work with real scenarios and real records — a rehearsal, not a lecture (Chapter 15). If your training opens with a tour of menus, you've already lost the adults in the room.

Earned through trusted messengers. Adoption travels peer to peer, not memo to inbox. People believe the colleague who says "this actually saved me an hour" over any executive announcement — which is why your champions come *from* the affected group, are named, are supported, and are visibly rewarded. The first adopters are being watched (Chapter 4). Make sure what happens to them is enviable.

Earned through visible response. When someone reports a problem and sees it fixed — and hears "here's what you told us, here's what changed" — adoption compounds. The feedback loop with visible fixes (Chapters 6 and 14) isn't customer service for your rollout. It's the fastest trust-builder you have, because it proves the thing announcements can't: that this time, someone is listening.

Earned through measurement. You cannot earn what you don't measure, and my story's organization measured nothing. Define adoption levels that mean something — aware, trained, *using*, *proficient*, *advocating* — and measure depth, not logins. A thousand people who logged in once is an announcement metric. Forty people who reach for the tool without being reminded, and two who proposed a new use for it — that's adoption, and Chapter 8 already gave you the scoreboard for it: safety to experiment, use cases spreading, time visibly reinvested, barriers dying young, ideas flowing uphill.

Do the five consistently and something shifts that you can feel in the hallways: the initiative stops being leadership's project that employees comply with, and becomes the team's tool that leadership supports. That's the verdict changing. That's earned.

One warning as you start: earning is slower than announcing — for about six weeks. Then it's dramatically faster, because announced adoption has to be re-announced forever (you've seen the organizations that relaunch the same tool every year), while earned adoption starts marketing itself. Seven years of announcements couldn't buy what one year of earning would have.

The next question is uncomfortable and unavoidable: who does the earning? Because your people won't render their verdict based on the champions or the training or the drumbeat alone. They'll render it based on the most visible evidence of all — what their leaders actually do.

That's Chapter 17.

The Executive Brief

Key Insight

Adoption is a verdict, not an event — your people judge everything you did before and after launch, and they judge with their behavior. It's earned through fingerprints, relevance, trusted messengers, visible response, and real measurement — never announced into existence. And internal adoption is the ceiling on customer adoption: your customers will never adopt what your people haven't.

Questions Every Leader Should Ask

- Which groups touched by our initiative were included in building it — and which were informed of it? Would they agree with our answer?
- Is our enablement problem-first (their time, their workflow) or feature-first (our menus, our modules)?
- Who are the trusted messengers inside the affected teams — and is what's happening to our first adopters enviable or cautionary?

- When someone reports a problem, how fast do they see a fix — and do they ever hear "here's what changed because you spoke up"?
- How do we measure adoption today: logins and completions, or depth — using, proficient, advocating?
- What are we currently reading as consent — tenure, silence, or the age of a promise — and what direct evidence of buy-in do we actually hold?

Monday Morning Actions

1. Find your excluded group: identify the largest team your initiative touches and audit their involvement honestly — designed with, or announced to? If it's the latter, schedule the design sessions this month. It is almost never too late to add fingerprints, and it is always expensive not to.

2. Rebuild your enablement around the capacity map: for the affected team's top ten time-consuming activities, show exactly which get lighter and how. Cut every training minute that doesn't map to one of them.

3. Stand up the adoption scoreboard with five levels — aware, trained, using, proficient, advocating — an owner, and a weekly review in the Chapter 14 rhythm. If your initiative is customer-facing, add the cascade metric: internal adoption reviewed *before* customer utilization, because the first number is the ceiling on the second.

Your AI Prompt

"I'm designing the adoption plan for an AI initiative affecting this team: [describe the team, their main responsibilities, and their ten most time-consuming activities]. Using adult learning principles, help me build: (1) a problem-first enablement outline that starts from their real work rather than tool features; (2) a champion structure — how many, selected how, supported how; (3) a feedback loop design with visible-fix communication; and (4) an adoption scoreboard measuring depth (using, proficient, advocating) rather than activity (logins, completions). Flag the three

most likely reasons this specific team would quietly decline to adopt."

Reflection

Think of a tool your organization launched that people still don't really use — the one that gets relaunched, re-trained, and re-announced every year or two. Now ask the verdict question: what evidence did the organization present to its people before and after that launch? Was their non-adoption irrational — or was it a correct reading of how they were treated?

Your 90-Day Checkpoint

By launch plus thirty days, you should be able to name: the affected groups with fingerprints on the design, the champions by name, the average time from reported problem to visible fix, and your adoption depth numbers — not logins. If your initiative is customer-facing, you should also be able to state your internal adoption level and say out loud: this is the ceiling on what our customers will do. If any of those are blank, stop announcing and start earning.

CHAPTER 17
Are Leaders Actually Leading?

The Natural Experiment

For a stretch of my career, I had a vantage point most transformation leaders never get.

I worked for an organization whose business *was* implementation — deploying our tool inside customer organizations and executing alongside them was the product. Which means I didn't watch one rollout at one company. I watched the same tool, implemented by the same team, using the same process, land inside organization after organization after organization.

Same software. Same methodology. Same implementers. Every variable held constant except one: the leadership on the other side of the table.

Scientists would call that a natural experiment. I called it an education — because the outcomes were not the same, and the difference announced itself early, in a pattern so consistent I could eventually predict an implementation's fate within the first few meetings.

In one kind of organization, the decision had been made at the top, and then people were *appointed*. Told what they'd be doing. And you could feel it in the room: the appointment was an inconvenience. Nobody had asked how this would fit their current workload — nobody had even *looked* at their current workload. Nobody had explained how this might improve their work. It was simply assumed, based on a title, that this person would be fine absorbing a major initiative on top of everything else. The initiative had been done *to* them, and they carried it exactly the way you'd carry anything strapped to your back without your consent: dutifully, minimally, and with no intention of going one step past the requirement.

In the other kind of organization, leadership did something that looked similar on an org chart and was completely different in the room. They found an *owner*. Not appointed — owned. And that owner became the source. They took pride in the thing. They asked us the right questions. They fed us the right information — the process realities, the exceptions, the data quirks — the things an implementation lives or dies on. They chased down answers inside their own organization. They were, in every meeting, the person who made the implementation solid.

Same tool. Same team. Opposite endings. When leaders appointed, we got burdened compliance and implementations that limped. When leaders created real ownership — clarified that someone was going to *own* the process rather than be *burdened* with it — you saw how beautiful adoption could be.

The technology was never the variable. The leadership was.

What Was Actually Happening

Put the two patterns under the light and the difference is precise:

Appointment transfers tasks. Ownership transfers meaning.

The appointed person received a workload. No why (Chapter 6). No workload math — nothing came off their plate to make room. No authority — they were coordinating someone else's decision. No stake in the outcome, because the outcome had never been connected to anything they cared about. Every signal said: *this is extra, this is yours to absorb, and it was decided without you.* Their minimal compliance wasn't a character flaw. It was — say it with me now — a correct reading of the situation.

The owner received something else entirely: a *charge*. Why this matters, why them specifically, what success looks like, what authority they carry, and what support stands behind them. Here's the thing about ownership that appointment can never replicate: **you can assign a task in a meeting, but ownership has to be accepted — and it's**

only accepted when it's worth accepting. The organizations that got beautiful adoption had made ownership worth accepting: visible authority, protected capacity, leadership backing, and pride available to whoever carried it well.

And notice what the owner did for the *partner* relationship — this is Chapter 13's collaborator principle from the other side of the table. As the implementing partner, we could only be as good as the information we received. Burdened appointees gave us the minimum, so we implemented against the official process (and you know from Chapter 10 what lives outside the official process). Proud owners gave us the truth — the exceptions, the workarounds, the political landmines — so we implemented against reality. Your partner's performance is, to a remarkable degree, a mirror of your owner's engagement. If your implementations keep disappointing across different vendors, look hard at what kind of owner you keep giving them.

So the question this chapter asks — *are leaders actually leading?* — comes down to something observable. Not sponsoring from a distance. Not approving and disappearing. During execution, leading means four visible behaviors: using the tools yourself where your people can see it (feet, not memos); showing up to the working sessions, especially the messy early ones; making ownership real — publicly transferring authority, not just responsibility; and killing barriers fast when your owner raises them, so that raising barriers stays worth doing (Chapter 8's blocker-age metric is your proof).

Your people can't see your intentions. They can only see your calendar.

Ownership Lives on the Street

Now, the question of *where* ownership should sit — because I watched organizations get this wrong in a specific, repeated way, and the wrong answer is usually the reflexive one.

The tool is technical, so it gets handed to IT.

Let me say this as plainly as I can: **IT is the last place ownership should live.** That is not a criticism of IT — it's a job description. IT is where *support* lives: governance, security, integration, data protection, uptime. Essential, involved from day one, non-negotiable. But support is not ownership, and confusing the two quietly kills business value.

If it's accounting software, it needs to be owned, operated, measured, and watched by *Accounting*. A customer service tool belongs to *Customer Service*. The owner needs to live on the street where the software operates — because only the person on that street feels the business consequences, sees the workflow reality, and watches the numbers that matter (Chapter 8's scoreboard belongs to them, not to a systems dashboard).

Here's what happens when IT owns a business tool: the system stays healthy and the value goes unmeasured. IT can tell you about uptime, tickets, and licenses — adoption theater's favorite metrics — because that's what's visible from their street. What IT cannot tell you is whether invoice accuracy moved, whether the onboarding got shorter, whether the bad-payer percentage shrank. Those numbers live where the work lives. Put the ownership there too, give IT the supporting role it's actually built for, and both functions get to succeed at their real jobs.

The Same Failure at Higher Altitude

Before I close this chapter, I want to point the lens up — because everything I've described about appointing versus owning happens at the very top of organizations too, and at that altitude it has a different name: the board mandate.

You've seen it, or you will: a board hands down the list. *Here are the twenty things this organization will accomplish over the next five to seven years.* AI adoption is on it, somewhere between a market expansion and a margin target. And the list arrives with no idea — none — of the conditions inside the organization. The resources. The processes. The people. Whether the operational alignment exists to execute even five of the twenty.

That's appointment at the highest altitude. The organization has been handed a workload with no workload math, told what it will do with no examination of what it *can* do. **A mandate without conditions is a wish with a deadline.** And executives receiving one face the same choice I gave transformation leaders in Chapter 9: negotiate the conditions before accepting the accountability — bring the board the readiness picture, the Scorecard profile, the honest sequence — or spend five years being measured against a list that was never connected to reality.

And one more widening of the aperture, because it needs saying at least once in this book: nothing in this chapter — nothing in this *book* — is actually about AI. Appointing instead of owning, mandating without conditions, preparing systems three times harder than people: these are leadership behaviors, and they show up wherever leaders drive change. Watch a merger or an acquisition sometime. The identical patterns appear — the silent cutovers, the excluded frontline, the assumed buy-in, the appointed integration leads — except the price isn't a failed system implementation. It's an entire acquired company's worth of value, trust, and talent. AI is simply where these behaviors are being exposed *right now*, at unprecedented speed. Fix the behaviors, and you've fixed more than your AI initiative. You've fixed how your organization changes.

That's why the leaders in my natural experiment who created real ownership didn't just get better implementations. They ran better organizations — and we could see it from the outside, one engagement at a time.

Leading, it turns out, is also a verdict. Your people render it weekly, based on your calendar, your usage, your owners, and your follow-through. The next chapter is about what they should see you doing with the tools themselves — because there's a way of using AI that leaders keep defaulting to, and it leaves most of the value on the table.

That's Chapter 18.

The Executive Brief

Key Insight

Appointment transfers tasks; ownership transfers meaning — and ownership must be worth accepting: authority, capacity, backing, and pride. Ownership lives on the street where the software operates — IT is the support system, not the address. And a board mandate without conditions is a wish with a deadline: the same appointing-without-conditions failure at the highest altitude.

Questions Every Leader Should Ask

- For our initiative: did we appoint someone, or did someone accept ownership? Would they use the same word we would?
- What came off our owner's plate when this went on it — and if the answer is nothing, what are we actually expecting?
- Does our owner live on the street where the tool operates — or did we hand a business tool to IT because it looked technical?
- What do our people see on our calendars: working sessions and personal tool use, or approvals and absence?
- If our board handed down its multi-year list today, could we answer with a conditions picture — or only with a salute?

Monday Morning Actions

1. Run the ownership conversation with your initiative's owner — as a conversation, not a confirmation. Five questions: Why does this matter? Why you? What comes off your plate? What authority do you carry, and who knows it? What does success look like for *you* if this goes well? If you can't answer all five, you appointed. Fix it this

week; it is rarely too late to convert an appointment into ownership.

2. Check every tool's address: list your active initiatives and their owners' home functions. Any business tool owned by IT gets a new owner on the street where it operates — with IT formally chartered in the support role it deserves.

3. Audit your own visibility: pull your last four weeks of calendar. Count working-session attendance and personal tool use your people could actually see. If the count is zero, your people have already counted it too.

Your AI Prompt

"I'm establishing ownership for an AI initiative involving [tool/process] in [function]. Help me: (1) draft the ownership charter — the why, the authority granted, what's removed from the owner's workload, the scoreboard they own, and the support IT provides in a defined supporting role; (2) script the public authority-transfer announcement so the organization hears ownership, not appointment; and (3) list the five ways this ownership structure is most likely to erode over 90 days, and an early-warning sign for each."

Reflection

Think of the best owner you've ever watched carry an initiative — the one who became the source. Now think about what their leadership did *before* they showed up that way: what was said, what was granted, what was removed, what was made safe. How much of what you admired in that person was actually created by the conditions around them? And who in your organization is currently carrying an appointment that the right conditions would turn into ownership?

Your 90-Day Checkpoint

Execution is being led — not just sponsored — when four things are observable: your owner would say "mine," not "assigned to me"; the tool's address matches its street; your leaders' calendars show working sessions and visible personal use; and barriers raised by the owner die within days, not weeks. Your people are already scoring all four. This checkpoint just asks you to see what they see.

CHAPTER 18
Stop Using AI Like a Tutor

Show Me Your Last Ten Prompts

There's a request I make when a leader tells me their organization is "using AI," and it cuts through every dashboard and adoption report in seconds:

Show me your last ten prompts.

Not the license count. Not the training completion rate. The actual prompts — what your people, and you, asked the machine to do this week. Because the prompts never lie, and in most organizations they tell the same story: *Explain this concept. Summarize this document. What does this term mean? Give me an overview of...*

Questions. Definitions. Summaries. Homework help.

So let me make a statement that sounds harsher than it is: if you're using AI mainly to solve mysteries and learn things you don't already know — and stopping there — you're using it wrong.

I know. AI is phenomenal at answering questions, explaining concepts, and helping you learn something new. Nothing about that is bad, and I'm not going to pretend otherwise. But if that's where your organization's use of AI begins and ends, you are leaving the majority of the value — the transformational majority — sitting on the table.

Here's the distinction, and it's the whole chapter:

Most people use AI like a tutor. The highest performers use it like a teammate.

A tutor answers your questions and hands the work back to you. A teammate takes on part of the work itself. The difference isn't the tool — same tool, same login, same monthly fee. The difference is what you're willing to *delegate*.

Don't just ask AI to explain a business strategy. Ask it to build one — then tear it apart with it.

Don't ask it to summarize the meeting. Ask it to identify the decisions, assign the actions, surface the risks nobody named, and draft the follow-up.

Don't ask it to teach you about leadership. Ask it to prepare you for tomorrow's difficult conversation — to challenge your assumptions, take the other person's side, and role-play the three ways the conversation could go.

Don't ask it what happened. Ask it what you should do next.

Feel the shift? Every pair starts in the same place — a real need — and the second version hands the machine actual work. That's the line. The moment you move from *"teach me something"* to *"help me accomplish something,"* you've crossed from using a search tool to working with a strategic partner.

And here's something you may not have noticed: you've been on the far side of that line for this entire book. Go back and look at the "Your AI Prompt" section in any chapter's Executive Brief. Not one of them asks AI to explain anything. They ask it to walk your problem upstream, stress-test your leadership structure, build your measurement plan, red-team your announcement, design your adoption scoreboard. Seventeen chapters of teammate prompts. If you've been running them, you've already felt the difference this chapter is naming.

What Was Actually Happening

Why do smart people — people who'd never dream of paying a brilliant new hire to do nothing but answer trivia — default to tutor mode with AI?

Three reasons, and none of them is a character flaw.

It's how we met these tools. The interface is a chat box, and a chat box invites questions. Twenty-five years of search engines trained us: type a query, get an answer, do the work yourself. We brought the search habit to a technology that isn't a search engine, the way early filmmakers pointed a movie camera at a stage play.

It's safe. A question risks nothing. Delegating real work — *draft this, build this, challenge me on this* — puts something of yours on the table and invites the machine into territory where you might be judged, or might have to admit the draft is better than your first attempt would have been. Tutor mode is safe precisely because nothing is delegated. That's also why it's cheap.

And nobody reframed the value. Most people think AI's greatest value is what it *knows*. Its greatest value is what it helps you *accomplish*. That reframe matters more every year, because the knowing gets commoditized — every organization's AI knows roughly the same things. Nobody builds an advantage on access to explanations. The organizations that win with AI won't be the ones asking the most questions. They'll be the ones redesigning how work gets done.

Which is why this sentence belongs in your next leadership meeting: **the real competitive advantage isn't knowing more. It's accomplishing more.**

Now connect this to what you already know. Chapter 4 taught you the AND — the deliberate pairing of what the machine carries and what your people keep. Tutor mode is the AND *unconsummated*: you bought the teammate and assigned it to answer trivia. The machine's actual strengths — volume, drafting, pattern-work, tireless iteration — only engage when real work crosses the desk. And remember my colleague from Chapter 5, the one saving a day a week? You don't get a day a week back by asking better questions. You get it back by handing over real work.

The Ladder

Usage depth climbs a ladder, and you can locate any person — or any organization — on it in about five minutes of prompt-reading.

Level one: Ask. Explain, define, summarize, look up. The tutor. Everyone starts here, and it's a fine place to start — it builds the fluency and trust the higher rungs require. It is a terrible place to stop.

Level two: Draft. First versions of real work products — the email, the analysis, the job description, the project plan, the board summary. The machine produces; the human shapes and decides. This is where time savings first become measurable.

Level three: Think. The rung most people never discover. *Challenge my assumptions. Argue the other side. What am I not seeing? Role-play my skeptical CFO. Stress-test this plan and tell me where it breaks.* Here the machine stops being an assistant and becomes a sparring partner — and the quality of your decisions starts to change, not just the speed of your output. Every prompt in this book's Executive Briefs lives on this rung or the next one.

Level four: Redesign. The work itself gets rebuilt around the pairing. Not "AI helps me do the old process faster" but "the process now looks different because the machine carries the volume and the human carries the judgment" — Chapter 4's task-by-task redesign, actually executed. Multi-step workflows, standing assignments, the machine woven into how the function operates rather than bolted onto how it used to.

Here's the organizational math that should get your attention: a workforce living on level one caps your AI return at rounding-error territory, no matter how high the license count climbs. The compounding returns — the day-a-week returns — live on levels three and four. Which means your job as a leader isn't to get more people using AI. It's to move the people already using it *up the ladder*.

Climbing It — Without Falling Off

Two things before you start pushing your organization upward, because this chapter needs its guardrails like every other.

First: **the teammate still doesn't get the judgment.** Delegating work is not delegating decisions. Chapter 4 drew this line and it doesn't move — accountability, relationships, and final calls stay human. Level-four organizations aren't the ones that removed people from the loop; they're the ones that put people at exactly the right points in it. The machine drafts. The person decides. Ask the machine for judgment and it will give you confident nonsense, faster than ever.

Second: **don't shame the tutor.** Level one is the on-ramp, and psychological safety (Chapter 3) applies to prompt quality too. If people get mocked for basic use, they don't climb the ladder — they climb off it, quietly, and you've manufactured non-adoption out of good intentions. The tone is always *"great, now let's go further,"* never *"you're doing it wrong."*

With those set, the climb is led the way everything in Part Four is led — visibly:

Model the rung you want. Share your own prompts, including the level-three ones where you asked the machine to challenge you. A CEO who shows the room "here's how I had AI red-team our pricing decision" moves more people up the ladder in five minutes than a training course does in a quarter — partly for the technique, mostly for the permission.

Celebrate depth, not volume. "Prompts run this month" is adoption theater (Chapter 8). The number worth celebrating is the reconciliation analyst who moved from summarizing reports to having the machine flag the anomalies and draft the exception queue. Put *that* story in the drumbeat.

Build the sharing ritual. A use-case library, a prompt-of-the-week, five minutes in the team meeting for "here's what I delegated this week." Depth spreads peer to peer, exactly like adoption itself (Chapter 16) — because the fastest way to discover the ladder's next rung is watching a colleague stand on it.

Do this, and the prompt audit that opened this chapter starts returning a different answer. Less *"explain this to me."* More *"build this with me."* And somewhere in that shift

— quietly, without a launch announcement — your organization stops using AI and starts working with it.

One more piece of Execute remains: the operating rhythm that keeps all of this alive week after week — the wins that compound, the barriers that die young, the cadence that outlasts the honeymoon. That's Chapter 19.

The Executive Brief

Key Insight

Most people use AI like a tutor; the highest performers use it like a teammate — and the difference isn't the tool, it's what you're willing to delegate. AI's greatest value isn't what it knows but what it helps you accomplish, and the returns compound as your people climb the ladder: Ask, Draft, Think, Redesign. Move from "teach me something" to "help me accomplish something," and a search tool becomes a strategic partner.

Questions Every Leader Should Ask

- If we pulled our organization's last hundred prompts, what would the ladder distribution look like — and what would mine look like?
- Where does our AI use stop: answers, drafts, thinking partnership, or redesigned work?
- Have we ever asked AI to challenge a decision we were excited about — and would we have listened?
- Are we celebrating usage volume, or usage depth? Which one do our success stories feature?
- Is it safe in our organization to share a prompt — including a basic one?

Monday Morning Actions

1. Run the prompt audit on yourself first: pull your own last ten prompts and place each on the ladder. Then, for one real task on this week's plate, deliberately climb one rung — if you'd have asked for a summary, ask for the decisions, risks, and a drafted follow-up instead.

2. Take your next significant decision to level three: give AI the context and ask it to argue against your position, surface what you're not seeing, and role-play the stakeholder most likely to object. Twenty minutes. Then notice whether the decision changed — or just got more defensible.

3. Start the sharing ritual: five minutes in your next leadership meeting for each executive to share one prompt they actually used. You'll accomplish three things at once — an instant depth audit, a permission signal, and the first entries in your use-case library.

Your AI Prompt

"Here is a piece of real work on my plate this week: [describe the task and its context]. Instead of explaining or summarizing anything, work it with me as a teammate: (1) produce a first draft or first version; (2) then switch sides — challenge the assumptions in your own draft, identify what's missing, and flag the risks I'm most likely to underestimate; (3) role-play the person most likely to push back on this, and make their strongest case; (4) then give me the revised version and the three decisions that remain mine to make. Those decisions are where I'll take over."

Reflection

Think about the best teammate you've ever had — not the one who answered your questions, but the one you could hand half-formed work to and get back something better, sharper, further along. Now be honest: have you ever given AI the chance to be that? Or have you been paying a teammate's price for a tutor's help?

Your 90-Day Checkpoint

By the end of this month, you should be able to point to: your own prompt history showing level-three use, one decision that changed because the machine challenged it, one workflow candidate for level-four redesign, and a sharing ritual that has met at least twice. If your organization's prompts still read like homework help, revisit the safety and modeling first — people climb ladders they've watched their leaders climb.

CHAPTER 19

Weekly Wins and Removing Barriers

Get the Water Out of the Boat

Two mandates from my career have stayed welded together in my memory, because they were the same mandate wearing two different problems.

The first involved invoicing. Roughly three-quarters of our invoicing was not going out on time — and if you know anything about accounts receivable, you already understand the blast radius of that number: cash flow you can't see, customers recorded as late payers who were never late, decisions made on receivables data nobody trusted. You've heard where that road ends; this book opened there.

The second involved our issue backlog. Old customer issues were sitting in the system — not hundreds, *tens of thousands* — some of them ancient, all of them still technically open.

In both cases, the mandate came down with real urgency, and in both cases it was the same mandate: *clean it up.* Get the invoices out. Get the old issues closed. Work the backlog down.

Get the water out of the boat.

And in both cases, the questions that actually mattered were never part of the assignment. For the issues: *why are they happening? Who are they happening to? What type of customer? What's the impact?* For the invoices: *why is three-quarters of our invoicing late — what upstream condition manufactures that number every single month?* Nobody asked. And — you could finish this sentence yourself by now — nobody was measuring whether any of the cleanup was changing the underlying rate at all.

So the bailing began. And here's the part that makes these stories belong in a chapter about execution rhythm: *the bailing had a rhythm.* There were updates. There was activity. There were numbers that moved — issues closed this week, invoices pushed out this week. If you'd audited the calendar, you'd have found something that looked exactly like disciplined weekly execution.

But the boat kept filling. Because the holes had no owner, no questions, and no place on anyone's agenda. We were executing — visibly, energetically, measurably — against the water level, while the leaks ran at their original rate, unexamined.

A rhythm of bailing is still drowning. It just has better attendance.

What Was Actually Happening

I want to give you the distinction those mandates were missing, because it's the difference between a weekly rhythm that transforms an organization and one that merely exhausts it:

Motion is not momentum.

Both look like execution. Both produce updates. Only one compounds. And the difference lives in what kind of win your rhythm is producing — because there are exactly two kinds.

Bailing wins reduce the accumulation. Issues closed, invoices pushed out, backlog worked down. They're real, they're countable, and they're deeply satisfying to report — which is precisely their danger. A bailing win *evaporates*: close five thousand old issues while the unexamined cause generates five thousand new ones, and your net progress is zero, delivered with excellent activity reports. Bailing restores capacity. It never changes the rate.

Plugging wins eliminate a cause. The contract language that guaranteed disputes — rewritten. The system handoff that made invoices late — fixed. The onboarding gap that manufactured a whole category of issues — closed. A plugging win doesn't evaporate. It compounds — every week after, forever, the boat takes on less water, whether anyone is bailing or not.

Why do organizations rhythmically bail and rarely plug? You already know the anatomy from Chapter 7 — the water is visible, countable, and has a report; the holes are quiet, cross-functional orphans. And you know from Chapter 8 that a backlog is a *lagging accumulation* — sediment from upstream causes — so working the backlog is managing the past while the present manufactures more of it. What these stories add is the execution-level consequence: **a weekly rhythm pointed at the water level will happily run forever.** Bailing generates its own justification — there's always more water — and the activity itself becomes the evidence that the team is executing. Entire departments have spent years in energetic, well-attended, thoroughly reported drowning.

Let me be fair, the way Chapter 7 was fair: sometimes you must bail. When the boat is at the gunwales — when cash is critical, when customers are furious — you bail, hard, and there's no shame in it. The discipline is the same one you learned for bridges and symptoms: *know which one you're doing, say it out loud, and never let bailing be the whole plan.* Bail with one hand. Plug with the other. And measure yourself on the leaks.

The Rhythm That Compounds

So here is the weekly rhythm I build now — the one those two mandates never got. It's one standing meeting, five questions, forty-five minutes, and it is the beating heart of your entire ninety days (Chapter 14 scheduled it; this is what happens inside it).

One: What moved on the leading indicators? Not activity — the scoreboard from Chapter 8, reviewed at the speed of the work. Rhythm first, financials after.

Two: What hole did we plug this week? The question that separates this rhythm from every status meeting you've ever sat in. Which *cause* died this week? Some weeks the honest answer is "none" — say it, and let the silence do its work. A rhythm that can go a month without plugging a single hole isn't executing. It's bailing with minutes.

Three: What did we learn? This is where the standing questions live, permanently — eight of them, in two sets. Four about the *leak*: why are these still happening, to whom, what type, what impact? And four about the *aim*: what does success actually look like, how is it measured, what are we really trying to accomplish, and are we solving a symptom or the source? The leak questions harvest intelligence — every week's bailing throws off root-cause data, *if the rhythm has a place to catch it*; the team closing old issues is sitting on the best diagnostic dataset in your company, and the mandate never asked them to look. Yours will. The aim questions guard against drift — because mandates mutate quietly, and a rhythm that stops re-asking what it's actually trying to accomplish will wake up one quarter having optimized the wrong thing beautifully.

Four: What's blocking us — and how old is the oldest barrier? More on this in a moment, because barrier removal is the leadership half of the rhythm.

Five: What win do we tell — and is it a bailing story or a plugging story? Wins that aren't told don't travel (Chapter 16), so every week feeds the drumbeat. But label them honestly, and let the plugging stories get the spotlight. What you celebrate is what you'll get more of — celebrate water removed and you'll get bailers; celebrate leaks killed and you'll get plumbers.

That's the whole machine. Five questions, weekly, same day, same people, leader present (Chapter 17 — your attendance *is* the message). It survives the honeymoon precisely because it's small: too light to skip, too pointed to fake.

Barriers Are Leadership's Inventory

Now the fourth question, expanded — because if wins are the team's output, barriers are *yours*.

Every initiative accumulates friction: the access nobody granted, the policy nobody updated, the decision stuck between two functions, the data feed nobody owns. Individually small. Collectively, they're how momentum dies — not in one dramatic failure, but in a quiet pile-up of unresolved little things, each one teaching your people that raising issues changes nothing.

So run barriers like inventory. Every barrier gets logged with three fields: the date it surfaced, the single named owner (Chapter 9's rule — never a committee), and the kill date. The metric that matters is the one Chapter 8 gave you: **the age of the oldest unresolved barrier.** Review it weekly, out loud, with the owner's name attached. Nothing measures your leadership's seriousness more honestly, because barrier age is really a trust metric: people only keep surfacing barriers while surfacing them visibly works (you learned that law in Chapter 5 — governance that doesn't move gets routed around, and so do leaders).

And when a barrier dies, *celebrate the kill like a win* — because it is one. Every barrier removed is friction deleted for everyone who comes after, which makes barrier removal the purest plugging win a leader can personally deliver. In my experience, the fastest way to convince a skeptical team that this time is different is embarrassingly simple: they raise a blocker on Tuesday, and it's dead by Friday, and someone says so out loud.

Closing Part Four

That's Execute — and look at the machine you've now built.

Adoption earned through fingerprints, relevance, messengers, response, and measurement (Chapter 16). Ownership that lives on the street, backed by leaders whose calendars match their announcements (Chapter 17). A workforce climbing from tutor to teammate (Chapter 18). And a weekly rhythm that compounds wins, catches intelligence, and keeps barriers young (this chapter) — a rhythm that bails when it must and plugs by design.

Run that machine for a quarter and you'll pass your day-90 gate with evidence instead of applause. But now comes the question that separates a successful initiative from a transformed organization: how do you make this permanent? How does one initiative's discipline become the way the whole organization operates — through the next tool, the next wave, the next decade?

That's Optimize. Turn the page one more time.

The Executive Brief

Key Insight

Motion is not momentum. Bailing wins reduce the backlog and evaporate; plugging wins eliminate causes and compound. A weekly rhythm pointed only at the water level will run forever and change nothing — so run five questions weekly: what moved, what hole died, what did we learn, how old is the oldest barrier, and what story do we tell. Bail when you must. Plug by design. Measure yourself on the leaks.

Questions Every Leader Should Ask

- Of the wins we celebrated last quarter, how many were bailing and how many were plugging — and would our team label them the same way?
- Do the people working our backlogs have a standing place to report what the backlog is teaching them about causes?
- What is the age of our oldest unresolved barrier — and does anyone in this organization know that number today?
- When someone raises a blocker, what's the average time until they see it die — and until they hear it acknowledged?
- Is there any cleanup mandate running right now with no paired mandate to find and fix the cause?

Monday Morning Actions

1. Audit your active mandates for boats: any directive that says "work it down" without asking why, to whom, what

type, and what impact is a bailing order. Pair each one, this week, with a named owner for the eight standing questions — four about the leak, four about the aim (what does success look like, how is it measured, what are we really trying to accomplish, symptom or source?) — and a leading indicator on the *rate*, not just the level.

2. Install the five-question weekly review on your initiative — forty-five minutes, same day, leader present. Run it three weeks before judging it; the first week is always awkward and the third is always worth it.

3. Open the barrier log: date surfaced, single named owner, kill date, reviewed weekly with the oldest barrier's age read aloud. Then kill one visible barrier within seven days and make sure the person who raised it hears about it. Trust compounds from exactly there.

Your AI Prompt

"Our team is working down a backlog of [describe: type, volume, age]. Act as an operations analyst: (1) design the intake analysis that would reveal why these keep occurring — the categories, customer types, and upstream triggers we should tag as we work each item; (2) define the two metrics that would tell us whether we're actually gaining — the inflow rate versus the clear rate — and what each combination of those numbers means; (3) propose three likely root-cause hypotheses to test first based on the pattern I described; and (4) design a one-page weekly review that reports water level, leak rate, holes plugged, and oldest open barrier."

Reflection

Somewhere in your organization right now, a hardworking team is bailing — energetically, visibly, with weekly updates — and privately wondering why leadership never asks about the holes. They already know where the leaks are. What has your rhythm been asking them for: their buckets, or their intelligence?

Your 90-Day Checkpoint

Part Four is complete when the machine runs without ceremony: the five-question review has met at least four consecutive weeks, at least one plugging win has been told through the drumbeat by the person who earned it, the barrier log exists with every entry owned and dated, and the oldest barrier is younger than fourteen days. If all four are true, your day-90 gate will take care of itself — and you're ready to make this permanent.

PART FIVE — OPTIMIZE

"How do we make it last?"

You've built awareness, diagnosed honestly, chosen well, and executed with rhythm. If you stopped here, you'd have something most organizations never achieve: one initiative, done right, with evidence.

And within two years, it would fade. The rhythm would thin out. The scorecard would gather dust. The next wave of technology would arrive and find your organization starting over — because an initiative, however successful, is an event. What lasts is something else.

Part Five is about that something else: turning one initiative's discipline into the way your organization permanently operates. These last three chapters are about health, scale, and the future — in that order, because that's the order they depend on each other.

Let's start with health.

CHAPTER 20

Transformation Is a Capability, Not a Project

The Metaphor I Give Boardrooms

When I sit down with a board or an executive team — especially at the beginning, before the plans and the numbers — there's a metaphor I almost always share. I share it because I believe, in my heart, that it rings true. And after all these years and all these organizations, I've never found a better way to say what I most need leaders to understand.

An organizational body is not that different from a human one.

A body is made of cells. An organization is made of people. Cells come together to form tissues — people come together to form teams. Tissues form organs — teams form departments. And organs come together into systems that must work in concert: the body has its circulatory, digestive, nervous, skeletal, and muscular systems; the organization has its finance, operations, sales, service, and technology. No system survives alone. The circulation feeds the muscles; the muscles move the body; every system's health depends on the others.

And at the top of all of it sits the head.

The brain doesn't just sit above the body — it *governs* it. Direction. Health. Maintenance. Attention. Every movement, down to the simplest — left leg, then right — begins as a signal from the head, carried down through the nervous system to the part that must act. The body goes where the head sends it, at the pace the head sets, with the coordination the head provides.

That's leadership. Your organization moves on the signals you send — and only on those signals. If the head's messages are clear and consistent, the body moves with coordination. If the signals are absent, garbled, or contradictory — one message this week, a different one next — the body stumbles, and no amount of strength in the individual limbs can fix it. You've seen that truth running through this entire book: the unexamined why (Chapter 1), the silent cutovers (Chapter 6), the mandates without conditions (Chapter 17). Those weren't strategy failures. They were nervous system failures.

Now, here's where the metaphor earns its seat in the boardroom.

When a body is sick, it produces symptoms: sneezing, coughing, a temperature. When an organization is sick, it produces symptoms too: turnover, customer churn, a sales number that won't move, backlogs that keep refilling, collections that keep growing. And in both cases, there are two very different ways to respond.

Think about why most cold medicines exist. Not to heal you — read the label; they don't claim to. They exist to *remove the inconvenience of the symptoms*. The cough quiets. The fever drops. You feel functional. And the illness underneath continues exactly as before.

There is nothing wrong with that for a cold, because a cold heals itself. But an organization's illnesses don't. And yet entire operating models are built on symptom-inconvenience-removal — you've watched it happen throughout this book. The collections tool that made the receivables symptom quieter while the disease manufactured new bad debt (Prologue). The bailing mandates that worked the backlog down while the leaks ran at their original rate (Chapter 19). Treatments that made the noise stop — and were celebrated precisely because the noise stopped.

If you operate your organization by removing the inconvenience of its symptoms, hear me clearly: you are not getting better. You are quieting the noise. And quieted noise behaves the way it does everywhere in nature. It's the volcano that seems calm right up until it erupts. It's the iceberg calving — piece after piece dropping quietly into the ocean, each one manageable, each one absorbed — until the wave comes.

A body whose symptoms are treated while its disease progresses doesn't stabilize. It stays sick, or it gets worse. An organization is no different: treat symptoms as inconveniences and organic growth stops being an option — you'll be lucky if the inorganic kind survives. Whether the strain comes from inside or from something as violent as a merger or acquisition, certain things must be in place for the body to stay healthy and maintained.

Without them, your organization isn't transforming. It's waiting to die.

What Was Actually Happening

I give boardrooms that metaphor because it reframes the question leaders think they're asking me.

They think the question is: *"Which project should we do?"* A project is a treatment — a prescription with a start date, an end date, a budget, and often an outside specialist. Take the course, finish the bottle, declare the episode over. Organizations run on treatments: the reorganization, the system implementation, the AI initiative. Some treatments are genuinely necessary. This book just walked you through running one well.

But nobody becomes healthy through treatments. Health isn't a procedure you undergo. It's a *capability* your body maintains — an immune system that detects, vital signs that warn, reflexes that respond, and practices that prevent. A healthy person isn't someone who survived an excellent surgery. It's someone whose everyday systems catch things early, respond fast, and keep the whole body fit enough that most surgeries never become necessary.

Transformation is exactly the same. It's not a project your organization completes. It's a capability your organization maintains. The project ends; the capability endures. The project has a team; the capability has a nervous system. The project gets funded once; the capability gets exercised forever.

And here's what I most want you to see, because it reframes everything you've done since page one: *you weren't running a project these ninety days. You were building the body's systems.*

Look at what you actually installed. The leading indicators and baselines from Chapter 8 — those are **vital signs**, so the organization detects illness in month two instead of at the autopsy. The fear conversations, the amnesty, the barrier log — those are **pain signals, honored**: a body that numbs its pain doesn't get healthier, it gets injured worse, and an organization that punishes bad news does exactly the same (you learned what silence costs in Chapters 3 and 5). The communication drumbeat — the **nervous system**, carrying clear signals from the head to every limb, and sensations back up. The weekly rhythm from Chapter 19 — the **heartbeat**: small, constant, unglamorous, and the difference between alive and not. The Scorecard — the **physical exam**, done honestly, on a schedule. The acknowledgment loop — the feedback that tells the brain whether the leg actually moved.

None of those are project artifacts. They're organs. And the question of this chapter is whether they keep functioning after the applause — because that's the difference between an organization that did a transformation and an organization that *has* transformation.

Keeping the Body Alive

So how do you keep the systems running when the initiative that built them is finished? Five commitments — and notice that not one of them is a new framework. This chapter closes the book's tools; from here we only make them permanent.

Give the capability an owner — a standing one. Projects have teams that disband. Capabilities need a home: someone who owns the health system itself — the rhythm, the scorecard cadence, the barrier log, the drumbeat — after the first initiative ends and independent of any single tool. In some organizations that's a transformation leader with Chapter 9's five conditions made permanent; in others it's an operating-rhythm ownership written into an executive's actual role. What it cannot be is "everyone," and it cannot dissolve with the project team. Bodies don't outsource their nervous systems.

Schedule the physicals. Re-score the Scorecard on a cadence — quarterly while you're building the capability, twice yearly once it's mature, and *always* before the next initiative and after any major organizational change. A merger or acquisition is major surgery on the organizational body; running one without re-examining the patient is how the identical failure patterns you've read in this book get repeated at ten times the cost. Compare every scorecard to the last one. Health is a trend line, not a snapshot.

Fund maintenance, not just treatments. Organizations budget generously for interventions and nothing for upkeep — then act surprised when the systems atrophy. The rhythm meetings, the champion development, the tool evaluations in the fast lane, the drumbeat, the periodic deep dives with the front line: these are the organization's diet and exercise. They're cheap. They're boring. They are the entire difference, and they need a permanent line, not a project code.

Prove it with a second cycle. Here's the honest test of whether you built a capability or just completed a project: *can you do it again, without heroics?* Take the next initiative — there's already a candidate list from your capacity map — and run the same loop: diagnose, charter, partner, rhythm, gates. If the second cycle runs on the systems instead of on individual heroism, the capability is real. If it requires rediscovering everything, you had a project with excellent PR. (And watch the second cycle move faster — that's Chapter 1's compounding, arriving on schedule.)

Keep the head connected. Last, and most important, because every system in the body answers to it: leadership attention is the one input the capability cannot survive without. The moment the head stops sending clear signals — stops attending the rhythm, stops asking for the vital signs, stops honoring the pain — the body doesn't collapse dramatically. It just begins, quietly, to get sick again, and the symptoms will be along eventually with their familiar inconvenience. Direction, health, maintenance, attention: those were the head's jobs on the first page of this chapter. They never stop being the head's jobs. That's not the burden of leadership.

That's the definition of it.

An organization with those five commitments doesn't dread the next wave of technology, the next market shift, or even the next acquisition — because it isn't betting on any single treatment. It's *fit*. And fitness, it turns out, scales — which raises the next question: when the first initiative succeeds and the whole organization wants what that team has, how do you grow it without breaking it?

That's Chapter 21.

The Executive Brief

Key Insight

An organization is a body: people are its cells, departments its organs, leadership its head — and every movement begins as a signal from the brain. Treating symptoms removes their inconvenience, not the illness; quieted noise erupts. Transformation is therefore a capability, not a project: vital signs, pain signals honored, a nervous system, a heartbeat, and scheduled physicals — owned permanently, funded as maintenance, and proven by a second cycle that runs without heroics.

Questions Every Leader Should Ask

- Which of our current "solutions" are cold medicine — removing the inconvenience of a symptom while the illness continues?
- What quieted noise are we currently mistaking for health — and what would the eruption look like?

- Who owns our transformation capability after this initiative ends — by name, in their role description, with conditions?
- Are our vital signs, heartbeat, and physicals funded as permanent maintenance, or coded to a project that's about to close?
- Could we run the next cycle tomorrow without heroics — and how do we know?

Monday Morning Actions

1. Run the cold-medicine audit: list your five most visible current problems, and for each, write what's being done and label it honestly — healing or noise-quieting. Any problem being quieted for the second time or more goes straight to the eight standing questions from Chapter 19.

2. Appoint the capability owner: name the person who owns the health system itself — rhythm, scorecard cadence, barrier log, drumbeat — beyond this initiative. Write it into their role, with Chapter 9's five conditions attached, and announce it the way Chapter 17 taught you to announce ownership.

3. Put the physicals on the corporate calendar for the next twenty-four months: quarterly re-scores this year, plus a standing trigger — any major organizational change, especially M&A activity, automatically schedules an exam before the ink dries.

Your AI Prompt

"Act as an organizational health advisor. Here are our most persistent visible problems: [list them — e.g., turnover in X, churn in Y, flat number in Z] and what we're currently doing about each. For every one: (1) classify our response as symptom-inconvenience-removal or root-cause treatment, and explain the tell; (2) name the likely underlying condition each symptom points to and which organizational 'system' it lives in; (3) tell me what 'quieted noise' exists here — where a previously loud problem went silent without the cause being addressed — and what the eruption scenario looks

like; (4) design the vital signs that would detect each condition recurring, months before the symptom returns."

Reflection

Think about how you treat your own health: the checkups you keep or skip, the symptoms you medicate versus investigate, the maintenance you fund versus defer. Now look at your organization's habits with fresh eyes. Are they different? Leaders tend to run their organizations the way they run their bodies — what is yours learning from you?

Your 90-Day Checkpoint

The capability exists — not just the project — when four things are true: a named owner holds the health system in their actual role; the physicals are on the calendar with an M&A trigger attached; maintenance has a permanent budget line; and the second cycle has a start date. If your initiative succeeded but those four are blank, you got a great treatment. This chapter is asking you to build a great body.

CHAPTER 21
Scaling and Governing AI

The Number Nobody Would Own

At one organization in my career, we had a piece of visibility most companies would pay dearly for: we could see, clearly, how many of our customers were *not* using our system.

Think about what that number represents. Every one of those customers had already bought. Already onboarded. Already represented the hardest-won kind of revenue there is. And a visible portion of them were not actually using what they'd purchased — which meant lower value realized, weaker renewals coming, and a quiet churn engine running under the surface.

Here's what we had built around that number: nothing. No strategy to invite customers into the system. No organization-wide approach to activation — not from the very beginning, not later. And no one measuring it as anything more than a fact that appeared on a report.

The growth was sitting right there. Visible. Unclaimed.

So I did what I do — I looked at that number strategically, because that is genuinely how I think, and I submitted a proposal to our CEO. It went something like this:

Make the number a lever, everywhere it touches the business. Build usage expectations into onboarding, so utilization starts high instead of being rescued later. Tie usage percentages to contract renewals, so the number has consequences. Stand up a task force whose entire job is going after non-usage. Train our customer success reps to *talk to the number* on every visit — so each visit becomes a statement of exactly where that customer stands, and the conversation can end only one of two ways: either we have an opportunity to grow them toward one hundred percent utilization, or they tell us what's wrong so we can improve it

— and not just fix it for them, but fix it for every customer who hits the same wall.

And then the piece I was most excited about: *bring in Finance.* Build revenue models around usage, so we could predict revenue outcomes from utilization movements — usage as a set of levers for organic growth, running right alongside the company's inorganic growth, while we simultaneously learned to onboard better so the percentages started high from day one. From there, the number becomes instrumentation: watch when accounts move above or below thresholds, watch when new users get introduced to the system — or don't — and treat every one of those movements as a trigger. A nervous system for the customer base, built from a number we already had.

The response?

Pushback. And the form it took has stayed with me ever since: *they didn't know if the numbers were true.*

Not "the strategy is wrong." Not "we can't afford the task force." The data itself — our own system's own usage data — was suddenly uncertain. And so the proposal, and the number, and the growth sitting inside it, stayed exactly where they were.

Unclaimed.

And the truth is, we never got there. Whether it was the sense that it would take too much time, or fear, or doubt, I can't tell you. What I can tell you is what it cost: the opportunity was unbelievable — the easiest lift we had — and a negotiating tool, sitting inside every renewal conversation we would ever have, that was never once used.

What Was Actually Happening

Two lessons live in that story, and together they are the whole discipline of scaling.

First: deployment is not scale. That organization had "scaled" its product in the way most organizations mean the word — sold widely, installed broadly, present everywhere. And underneath the deployment sat the truth: a visible fraction of it was actually being *used*. The gap between installed and utilized is one of the most consistently ignored numbers in business, and it's where growth hides in plain sight — because expanding utilization of what's already deployed is the cheapest growth any organization will ever buy. No new sale. No new build. Just the distance between what you have and what you use.

Hold that thought, because it's about to become your AI scaling strategy.

Second: doubting the number is cheaper than owning it. Notice what "we don't know if the numbers are true" actually accomplished. If the number was true, someone would have to own it — the task force would need standing up, the renewal conversations would get harder, the onboarding would need redesign, and a visible metric would start measuring people's performance. Doubting the data made all of that optional. Data doubt is the last refuge of the status quo: it sounds like rigor, costs nothing, and conveniently postpones everything. You've seen its cousins throughout this book — the two years of "vetting" (Chapter 2), the benchmarks never requested again (Chapter 9). The healthy response to an untrusted number isn't to shelve it. It's to *govern* it: give it one definition, one owner, and a validation path — and then act, because now nobody gets to hide behind uncertainty you've eliminated.

And there's one more thing that story crystallized for me — about my own profession this time.

A strong transformation leader doesn't look at people as *the* lever, or systems as *the* lever. They look at the organization holistically — the way a chemist reads a reaction, or the way someone plays a very complicated strategic board: every element in relation to every other, every move rippling somewhere it wasn't aimed. And here is what most executives get wrong about what we do: they assume transformation is mostly *elimination* — cutting cost, removing steps, reducing heads. Elimination has its place; it's a natural part of the process. But it was never the point.

The real work of transformation is increase. Finding the levers that grow what you already have — utilization, capacity, retention, renewal — because moving organic growth is so much less expensive than buying the inorganic kind. A good transformation leader truly understands those levers, all of them, and reaches for increase first. That proposal was me pulling one of those levers.

The organization declined the pull. Yours doesn't have to.

Keep all of it in hand. Because you are about to be the organization in this story.

Scaling: Depth Before Breadth

Here's the moment this chapter exists for. Your first initiative worked. The day-90 gate passed on evidence. And now everyone wants what that team has — the board is asking "where else?", the vendors are circling, and the temptation is to do what most organizations do: *deploy everywhere*. Licenses for everyone. The tool in every department by Q3.

That's how you build my story's company: broadly installed, thinly used, with the gap invisible until the renewal — except your renewal is the next budget cycle, when someone asks what all those licenses actually changed.

So scale the other way. **Depth before breadth.** Your first expansion target isn't a new department — it's the utilization gap inside what you've already deployed: the winning team climbing the ladder from Chapter 18, the adjacent workflows in the same process, the adoption depth moving from *using* to *proficient* to *advocating* (Chapter 16). Wring the value out of what's installed, publicly, and you've done two things at once — compounded your evidence and taught the organization that here, deployment is the beginning of an initiative, not the end of one.

Then, when you do spread — and you will — you scale the *system*, not just the software. Everything from my CEO proposal, turned inward:

Instrument everything. Adoption depth per team, visible on one page. Thresholds and triggers: an alert when a team's usage drops, when a new hire's first month passes without the tools, when usage plateaus below proficiency. Chapter 20 called the drumbeat your nervous system; this is its sensory layer — nerve endings in every limb. You cannot scale what you cannot feel.

Talk to the number. Train your leaders and champions to do with internal teams exactly what I wanted our reps doing with customers: every low-usage conversation ends one of two ways — an expansion opportunity, or intelligence about what's broken, fixed once for everyone. Both outcomes pay. A usage report that nobody talks to is my story's report: a fact, visible and unclaimed.

Connect usage to money. Bring Finance in early, exactly as I proposed. Model the chain — adoption depth to time saved to capacity reinvested to business outcome — so expansion decisions ride on predicted, then proven, value rather than enthusiasm. When Finance can forecast what moving a department from *using* to *proficient* is worth, scaling stops being a faith initiative and becomes a portfolio with returns. That's how organic AI value grows alongside whatever else your business is doing.

Onboard high. Every new team enters through the playbook you've already built — charter, champions, rehearsal, floor support, day-one measurement — so their usage starts high instead of being remediated later. This is the compounding Chapter 1 promised and Chapter 20 tested: the second team's launch should cost half the first one's, and the fifth should barely feel like a launch at all. If every expansion still requires heroics, you're not scaling. You're relaunching.

Governing: The Rules That Make Speed Safe

Scaling multiplies everything — including everything you'd rather not multiply. Governance is how you scale without scaling your problems, and by now you know I don't mean a committee that says no slowly. Good governance is what makes speed *safe*. Four standing structures:

Govern the numbers first. One definition of every scaling metric, one named owner, a validation path anyone can inspect (you learned what unshared meaning costs in Chapter 10). This is the structural answer to "we don't know if the numbers are true" — the sentence should be *impossible to say* in your organization, because the numbers have owners and the owners have receipts. If you don't govern your numbers, your numbers can't govern your growth.

Keep the fast lane open. As AI spreads, tool requests multiply — and Chapter 5's law scales right along with them: governance that moves at the speed of the work gets used; governance that doesn't gets routed around, now by ten teams instead of one. Weeks not quarters, every no with a reason and an alternative, forever.

Gate every expansion with the same discipline. New initiative, new department, new tool — same seven-question filter (Chapter 12), same confusion test (Chapter 10), same charter, same scoreboard. Scaling isn't permission to skip the diagnostics; it's proof they work. The moment "we've done this before" becomes a reason to skip the gates is the moment you start scaling confusion at machine speed.

Protect the pairing. As AI touches more work, re-ask Chapter 4's question in every new territory: what must stay human here — which judgment points, which relationships, which accountability? Scale has a gravity toward automating everything that moves; your governance holds the line that replacement remains a design choice, made deliberately, visible to the people it affects. That line, kept at scale, is what your workforce is watching to decide whether the AND was ever real.

And over all of it: the physicals. Re-score the Scorecard before every major expansion wave (Chapter 20 already put it on your calendar). Scaling on a stale readiness picture is how healthy first initiatives become unhealthy fifth ones.

Do all this, and something happens that my story's organization never experienced: the number gets claimed. The visible gap between installed and utilized becomes somebody's job, with instruments, conversations, financial models, and triggers wrapped around it. Growth stops hiding in plain sight — because your organization has finally built the thing that was missing that day in the CEO's office.

Not better data. The willingness to be governed by it.

One chapter remains. You have the health (Chapter 20) and the scale (this chapter). What's left is the horizon — because the tools you've deployed will be obsolete sooner than you think, the next wave is already forming, and everything this book built was actually preparation for a question bigger than AI.

Turn the page one last time.

The Executive Brief

Key Insight

Deployment is not scale — the gap between installed and utilized is where growth hides, visible and unclaimed. Scale depth before breadth, instrument everything, talk to the number (every low-usage conversation ends in expansion or intelligence — both pay), connect usage to money with Finance, and onboard every new team high. And govern the numbers first: "we don't know if the data is true" is the last refuge of the status quo, and it should be impossible to say in your organization.

Questions Every Leader Should Ask

- What is our installed-versus-utilized gap right now — for our AI tools, and honestly, for everything else we've deployed?
- Who owns our usage numbers — one definition, one name, a validation path — and could anyone credibly say "we don't know if they're true"?

- When usage drops, plateaus, or a new hire's first month passes without the tools — does anything trigger, and does anyone act?
- Can Finance model what moving a team from using to proficient is worth — and are our expansion decisions riding on that math or on enthusiasm?
- Does our fifth launch cost less than our first — or are we relaunching with heroics every time and calling it scale?

Monday Morning Actions

1. Claim your number: pull the installed-versus-utilized picture for your AI deployment (and while you're at it, your customer-facing systems — my story is probably your story too). Give the gap one owner, one definition, and a place in the weekly rhythm.

2. Build the trigger list: define the three usage movements that demand a conversation — the drop, the plateau, the absent new user — and script the two-outcome conversation for each: expansion, or intelligence we fix for everyone.

3. Book the Finance session: one working meeting to draft the usage-to-value chain for your first initiative — adoption depth, time saved, capacity reinvested, outcome moved — and a first model of what the next expansion wave should be worth. Bring your Chapter 8 baseline; it's the seed data.

Your AI Prompt

"Here is our AI deployment picture: [teams deployed to, adoption depth levels per team, usage trends if known] and our first initiative's measured results: [baseline vs. current]. Act as a scaling strategist: (1) identify our installed-versus-utilized gaps and rank them as expansion targets by value and ease; (2) design the usage instrumentation — the per-team metrics, thresholds, and triggers we should monitor weekly; (3) draft the usage-to-revenue/value model I should build with Finance, including the levers and what each is

plausibly worth; (4) sequence the next two expansion waves, and tell me which gate from our playbook each wave is most at risk of skipping."

Reflection

Somewhere in your organization there is a number like mine — visible, reported, and unclaimed — showing exactly where value is leaking or growth is waiting. You may have already proposed doing something about it, and you may have already heard the pushback. Here's the question worth sitting with: was the data really in doubt? Or was the data the only thing in the room that couldn't defend itself?

Your 90-Day Checkpoint

You are scaling — not just spreading — when four things are true: the installed-versus-utilized gap has an owner and appears in the weekly rhythm; usage triggers exist and provoke actual conversations with two possible endings; Finance holds a working usage-to-value model that expansion decisions actually cite; and your most recent launch cost measurably less than your first. If deployment is growing while those stand still, you're building my story's company — and the renewal is coming.

CHAPTER 22

Ready for What's Next, Not Just What's Now

The Last Visit

I keep promising you we've been to that boardroom for the last time. This time I mean it.

You know the room by heart now. The receivables crisis on the screen. The vendor suggestion — *"a tool that could help us with collections"* — and the heads nodding around the table. The applause, weeks later, for six figures recovered. The question nobody asked. The benchmarks nobody requested again. You've stood in that room with me five times across this book, and each visit showed you one layer: what happened, the diagnosis error, the measurement error, my seat, and once — in Chapter 12 — a single decision rewound.

This final visit is different. This time, walk in carrying everything.

Walk in with Chapter 1, and notice there was never an examined why in that organization — AI had arrived like weather, and the only articulated reason was one nobody would have said to the workforce. Walk in with Chapter 3, and feel the fear in the building that no one had ever named, converting quietly into silence and shadow. Walk in with Chapter 7, and watch the room aim at the place where pain became visible while the problem was manufactured years upstream. With Chapter 8, and see the celebration of a lagging number while every metric that mattered went unexamined. With Chapter 9, and see the person carrying the evidence — the person whose job was to see — sitting without sponsorship, platform, or a follow-up loop. With Chapter 10, and know that even the tool they bought was pointed at data and processes no two departments described the same way. With Chapters 12 through 19, and recognize everything that was never built: no charter, no honest label, no gates, no prepared people, no earned adoption, no weekly rhythm asking what hole we plugged. And with Chapters 20 and 21, the deepest cut: no health system to catch it, and a leadership team reaching for elimination while the increase levers hung unpulled on the wall.

Every chapter of this book was in that room. As an absence.

Now run the room forward the other way — because you've earned the right to see the ending they never got. An examined why, safe to say out loud. Fears named in the team's own words, months before any tool. The problem walked upstream to onboarding and contracts and invoicing. A baseline dated at the source. A transformation leader with five conditions and a standing platform. A first initiative chosen by the filter — invoice accuracy, chartered on one page, staffed by the willing. A collaborator who got *more* interested when shown the mess. Ninety days, three gates, a weekly rhythm plugging holes. And then a meeting with applause in it — for the inflow to collections shrinking, for a problem being manufactured slower, for a leadership team reading its own vital signs and trusting its own aim.

Same organization. Same people. Same intelligence in the room. The only variable — say it with me one last time — was the conditions.

Here's the part I've saved for the end, and I'd give a great deal for it not to be true: I told you in Chapter 9 that I knew, with certainty, they would be back in that same room. They were. Nothing upstream had changed, so the pipeline kept its promise. That return trip — the one this entire book exists to spare you — is what an organization pays when it medicates instead of heals.

Symptoms and Diagnoses

Which brings me back, one final time, to the body.

You'll remember the metaphor from Chapter 20 — the cells and tissues and organs, the head that governs, the cough medicine that never claimed to heal. I'm returning to it here because *that boardroom is where the metaphor hits home*. Look at what was actually on the table that day.

The collections number was a *symptom*. A fever, spiking on a monthly report. The tool was *cough medicine* — it made the symptom quieter, genuinely and measurably quieter, and the room celebrated the quiet as if quiet were health. And the benchmarks I carried into that room — the aging trends, the invoice accuracy data, the upstream evidence?

That was a *diagnosis*. The room was looking at symptoms. A diagnosis had been brought to the table. And the organization chose the medicine that removes the inconvenience over the treatment that removes the disease — because suppression is fast, and diagnosis asks something of you.

So here is my charge to you, as directly as I can make it, because if you take one instruction out of this entire book, let it be this one:

Stop suppressing what is happening in your organization, and start treating the root causes.

I won't pretend it's the quicker path. It isn't. Suppressing the symptom is always faster — that's its entire sales pitch, in medicine cabinets and vendor demos alike. But treatment costs far less in the end, for one reason that should now be written somewhere you can see it: **suppression minimizes. Treatment eliminates.** The suppressed problem returns on schedule, invoice after invoice, quarter after quarter, forever — with interest. The treated problem is gone, and every week after, the boat takes on less water whether anyone is bailing or not. You are not choosing between fast and slow. You are choosing between renting your problems and ending them.

Ready for What, Exactly?

And now the title of this final chapter, because it carries the book's last secret.

Everything in these pages used AI as its occasion. But you may have noticed — I hope you've noticed — that almost nothing in it was actually *about* AI. The tools named in my stories will age. The platforms being demoed to you this quarter will be museum pieces in five years. The wave after this one is already forming, and I don't know its shape any more than you do — nobody honest does.

Here's what I know instead. The next wave will arrive exactly the way this one did: borrowed urgency in the boardroom, fear in the hallways, vendors selling to your loudest pain, and a choice between suppression and diagnosis. And when it arrives, the organizations that thrive won't be the ones that picked the right tool this time. They'll be the ones that built the *conditions* — the examined why, the honest conversations, the diagnostic lenses, the measurement discipline, the conditioned leaders, the heartbeat rhythm, the health systems. Those don't expire when the technology does. They are wave-proof, because they were never about the wave.

They're also bigger than technology altogether. The next test of your organization's conditions may not be a tool at all. It may be a merger — and I've told you what the identical failure patterns cost at that scale. It may be a market shift, a crisis, a sudden opportunity that rewards the organization that can run an honest ninety-day cycle and punishes the one still in year two of talking. "AI Ready," it turns out, was never really the destination. It was the training ground. An organization with these conditions isn't AI ready.

It's *change* ready. And change is the only thing I can promise you is coming.

Across the Table

So let me say the last thing the way I'd say it if we were sitting across from each other, the way this book has tried to sit with you all along.

Somewhere ahead of you is a boardroom moment. Maybe soon. The numbers will be sobering, the suggestion will be concrete and exciting, and the heads will start to nod — because that's what heads do when a symptom meets a salve. And in that moment, you will be something that room I sat in never had: a person carrying the diagnosis, *with the conditions for it to matter*. You'll know the questions now. *If this works perfectly, what's still broken? Where is this outcome created? What does success look like, and how will we know? Symptom — or source?*

Ask them. Out loud. That's the whole job.

You will not do all of this perfectly — I didn't, and I've shown you my receipts. Progress, not perfection, was the deal from the introduction, and it's the deal to the last page. But an organization that asks the right questions, honors its pain signals, treats its diseases, and pulls the increase levers is an organization that gets healthier every cycle — while the ones around it keep celebrating quieter symptoms in rooms that never change.

AI will never outperform the conditions of the organization that implements it.

You now know how to build conditions worth implementing anything in.

Go build them.

The Executive Brief

Key Insight

Every failure in this book's opening boardroom was a condition this book taught you to build — the room was looking at symptoms while a diagnosis sat on the table, and it chose the medicine that quiets over the treatment that cures. Suppression minimizes; treatment eliminates. And the conditions you've built were never about AI: they are wave-proof, merger-proof, crisis-proof. AI ready was the training ground. Change ready is the prize.

Questions Every Leader Should Ask

- Where in our organization right now is a symptom being suppressed and celebrated — and what would the diagnosis on that table actually say?
- If the next wave of technology arrived tomorrow, which of our conditions would meet it — and which of

this book's chapters would still be an absence in our room?

- Who in our organization is currently carrying a diagnosis we haven't created the conditions to hear?
- Which problems are we renting — paying for their return every quarter — that treatment would end?
- When the heads start nodding in our next boardroom moment, who will ask the questions out loud? Have we made it safe to be that person?

Monday Morning Actions

1. Run your own last visit: take your organization's most persistent, expensive problem into a room with your leadership team and walk it through this book's lenses — the why, the fear, the source, the scoreboards, the conditions, the process, the rhythm. Mark each one honestly: present, or absent. The absences are your map.

2. Choose one suppressed symptom and commission its diagnosis this month — the upstream walk, the eight standing questions, a named owner, and a date the treatment plan is due. Say out loud, to the team, which one you're doing: quieting or curing.

3. Put this book's system on next year's calendar before you close it: the physicals, the gates, the rhythm, the re-score — so that what you built this quarter is still alive when the next wave, whatever it is, arrives.

Your AI Prompt

"Here is my organization's most persistent problem and everything we've thrown at it over the years: [describe the problem and the history of attempted fixes]. Audit our history: (1) classify each past attempt as symptom suppression or root-cause treatment, with the tell; (2) estimate what we've paid in 'rent' — the recurring cost of this problem returning after each suppression; (3) lay out what a genuine diagnosis would require — the upstream walk, the data, the people to interview; and (4) tell me what conditions my organization would need for that diagnosis to

actually change the decision this time. Be blunt. I've read the book."

Reflection

You've been in the boardroom now — mine, and by the end of this book, your own. One question remains, and only you can answer it: the next time the heads start nodding, which person in that room will you be — the one nodding, the one silently knowing, or the one who asks? Everything in these pages was written so you could choose the third.

Your 90-Day Checkpoint

The book is complete when its system is running without it: the why said out loud, the fears named, the scoreboards live, the leaders conditioned, the rhythm beating, the physicals scheduled, and one problem — at least one — being treated at its source instead of suppressed at its symptom. If that's true, you're not just AI ready. You're ready for what's next. And what's next has no idea what's waiting for it.

CONCLUSION
A Letter Before You Go

Before you close this book, I want a few minutes with you — not as an author, and not as a consultant. Just as the person who sat in these rooms.

I need you to know what this book was never meant to do: place blame. If you've read these pages carefully, you may have noticed there are no villains in them — not the leaders who said "no one is ready," not the boards with their lists, not the vendors with their demos, not even the room that applauded the wrong victory. Every room in this book was full of capable, intelligent people doing the best they could with the conditions they had. Pointing fingers was never the point. It never is.

This book exists for a much simpler reason: I actually care.

I care about the people we serve. I care about the stakeholders we represent. I care about the organizations trying to live through one of the biggest changes our working world has ever seen. And I didn't write about these failures from the outside looking in — I sat inside these organizations. Different industries, different projects, different decades of technology, and I watched the same things happen again and again. Eventually I had to admit they all happen for one reason.

Organizations have countless variables. But there is one common denominator, everywhere, in every industry, in every project I have ever touched: *people*. People who need resources. People who need processes. People who need to engage with the technology in order for it to work for the organization at all. And people we cannot control — no matter what the org chart implies.

And if we're being honest? I'm not sure the machines are so different. We tell them what we want based on the outcomes we're hoping for, without ever understanding the journey it takes to get there. Sound familiar? It should. It's this whole book in one sentence.

But that's enough of me and my soapbox. Let me leave you with the part that matters.

Each of you carries a responsibility — to the people you serve and to the stakeholders who trust you. That responsibility doesn't care whether the change in front of you is AI, a merger, a new system, or whatever wave comes next. It only asks one thing of you: figure out what it takes. If it's not this book, then let it be another book, another teacher, another honest look in the mirror — I genuinely don't mind which. What I mind is the alternative: another room full of nodding heads, another quieted symptom, another workforce learning what silence teaches.

Because everyone deserves better. The people doing the work deserve better. The customers on the other end of every broken process deserve better. The stakeholders who trusted you with their investment deserve better.

And you — carrying the weight of all of it — you deserve better too. You deserve to lead an organization whose conditions are worthy of your effort.

Now you know how to build one.

Thank you for spending these pages with me.

With gratitude,

Valetta Taylor

If this book helped you, the journey continues: the AI Ready Scorecard™ worksheet in the Appendix is yours to use and reuse, and the companion edition is available at ecfsystem.com/aiready. The next books in this series — The Executive AI Playbook™ and Executive AI Prompts™ — pick up exactly where this one leaves off: with you, ready.

APPENDIX

Part One: The Twelve AI Ready Principles™

The ideas this book was built on — one page, for your wall, your wallet, or your next boardroom.

1. **AI will never outperform the conditions of the organization that implements it.** *(The thesis — every page of this book)*

2. **You cannot lead what you do not understand.** *(Chapter 2)*

3. **People don't resist AI — they resist uncertainty.** *(Chapter 3)*

4. **Replacement is a design choice, not a destiny.** *(Chapter 4)*

5. **Silence is also a communication.** *(Chapter 6)*

6. **Solve where value is created — not where pain becomes visible.** *(Chapter 7)*

7. **Measure what creates the outcome — not just the outcome itself.** *(Chapter 8)*

8. **Organizations don't fail for lack of capable people — they fail when capable people lack the conditions transformation requires.** *(Chapter 9)*

9. **Never automate confusion. Create clarity first.** *(Chapter 10)*

10. **The system does what we tell it to do; the people do what they have to do.** *(Chapter 15)*

11. **Adoption is earned — not announced.** *(Chapter 16)*

12. **The real work of transformation is increase, not elimination.** *(Chapter 21)*

Part Two: The AI Ready Scorecard™ Worksheet

A printable edition of the instrument from Chapter 11. Photocopy freely, or download the full-size companion edition at ecfsystem.com/aiready.

How to use it: Each executive scores independently — no discussion first. Rate every statement 1 (strongly disagree) to 5 (strongly agree). Score what your organization demonstrates, not what it intends: a 4 or 5 requires evidence from the last ninety days. If the evidence is "we're planning to," the score is a 2. Then compare results side by side — the variance between scorers is data.

Scorer: ____________________ **Date:** ___________

Dimension 1: Awareness Readiness

Do our people understand why AI matters — and is it safe to talk about?

#	Statement	Score (1-5)
1	Our executive team has communicated a why for AI that is anchored in our business and safe to say to our entire workforce.	___
2	We have held honest conversations about AI fears, and I can name our people's top concerns in their own words.	___
3	Our people hear about changes from leadership before they hear about them from the rumor mill.	___
4	We know where AI is already in use in our organization — sanctioned and unsanctioned.	___

5	A recurring communication drumbeat about our AI plans is scheduled — before, during, and after.	___

Dimension total: ___ / 25 — *If low: Chapters 1, 3, 5, 6*

Dimension 2: Strategic Readiness

Are we solving the right problem?

#	Statement	Score (1-5)
1	We have defined the specific business problem our AI initiative will address — in one sentence, consistently, across the executive team.	___
2	We have walked that problem upstream and can state where the outcome is created, not just where it becomes visible.	___
3	If our proposed initiative works perfectly, we can articulate what will still be broken — and we've labeled it honestly: source, bridge, or symptom.	___
4	Our initiative was selected by business value, not by which vendor demo matched our loudest pain.	___
5	We can say why this	___

	problem, why AI, and why now — without referencing what competitors are doing.	

Dimension total: ___ / 25 — *If low: Chapters 1, 7*

Dimension 3: Cultural Readiness

Will our culture support adoption?

#	Statement	Score (1-5)
1	It is demonstrably safe to be a beginner here — people ask basic AI questions in the open, not in private.	___
2	Employees can admit to using unsanctioned tools without fear of punishment, and some have.	___
3	Our people would describe our AI story as "people AND the machine" — and could point to leadership behavior that proves it.	___
4	When change comes, our people extend trust — we are not still paying interest on past rollouts done in silence.	___
5	Someone recently challenged a direction leadership was excited about, and the challenge changed	___

	the outcome.	

Dimension total: ___ / 25 — *If low: Chapters 3, 4, 6, 9*

Dimension 4: Leadership Readiness

Do we have the leadership this transformation requires?

#	Statement	Score (1-5)
1	Our initiative has an active sponsor who shows up, uses the tools, and spends political capital — not just approvers who signed off.	___
2	Our transformation leader has all five conditions: sponsorship, platform, visibility, cultural safety, and an acknowledgment loop.	___
3	Adoption, communication, and barrier removal each have a single named owner — none of them a committee.	___
4	Every executive can state the initiative's why in one sentence, and the sentences match.	___
5	When someone flags a problem, our organization returns to them with evidence of whether it was solved.	___

Dimension total: ___ / 25 — *If low: Chapter 9 — and be honest*

Dimension 5: Operational Readiness

Is the process we're about to touch actually ready?

#	Statement	Score (1-5)
1	The process we intend to automate passes the confusion test: people who do the work describe it consistently.	___
2	We have inventoried the workarounds — the manual fixes, re-keys, and translations quietly patching this process — before automation removes them.	___
3	The documented process matches the real one, or we know exactly where it doesn't.	___
4	We are automating at or near the source of the problem, not industrializing the cleanup.	___
5	We know which judgment points in this work must remain human, task by task.	___

Dimension total: ___ / 25 — *If low: Chapters 7, 10*

Dimension 6: Data & Technology Readiness

Can our data and systems support what we're asking of them?

#	Statement	Score (1-5)
1	In the process we're targeting, key terms and fields mean one thing — across every unit that touches them.	___
2	We know where our data degrades — every handoff where information gets re-keyed, guessed at, or lost.	___
3	The data this AI will learn from is reliable enough that we would defend decisions made on it today.	___
4	Our systems can integrate with what we're planning to deploy, and someone technical has verified that — recently.	___
5	Security, governance, and data protection have been evaluated for this specific use — including where our data goes and what the vendor's terms actually say.	___

Dimension total: ___ / 25 — *If low: Chapter 10*

Dimension 7: Measurement Readiness

Will we know if we're succeeding — in time to act?

#	Statement	Score (1-5)

1	We have captured a dated baseline at the source of the problem — not just the symptom.	___
2	We have both scoreboards defined: leading indicators for the first sixty days, and lagging indicators that prove business impact.	___
3	Each number has an owner and a review cadence, and rhythm gets reviewed at the speed of the work — not quarterly.	___
4	Our pivot triggers are written down — we agreed in advance what result would make us change course.	___
5	Our AI metrics measure business change, not adoption theater — we could distinguish a real win from a well-attended training.	___

Dimension total: ___ / 25 — *If low: Chapter 8. Do not spend until this is fixed*

Dimension 8: Implementation Readiness

Can we execute this successfully?

#	Statement	Score (1-5)
1	We have selected a first initiative that is	___

	meaningful enough to matter and small enough to learn from.	
2	We know what we need in an implementation partner, and how we'll evaluate one beyond the demo.	___
3	Change management — communication, training, inclusion of the people who do the work — is inside the implementation plan, not alongside it.	___
4	The people this change touches have been part of designing it, not just informed of it.	___
5	We have capacity to do this — the humans involved have hours, not just enthusiasm.	___

Dimension total: ___ / 25 — *If low: Chapters 12–15*

Dimension 9: Sustainability Readiness

Will this become a capability — or a project that fades?

#	Statement	Score (1-5)
1	We have a plan for what happens after launch: who governs, who improves, who evaluates the next use case.	___
2	Saved time and early wins have a defined reinvestment path —	___

	we can say where freed capacity will go.	
3	There is a standing, fast-moving channel for employees to propose new tools and uses.	___
4	Success stories move between teams by design, not by accident.	___
5	We treat transformation as a capability we're building — not an initiative we're completing.	___

Dimension total: ___ / 25 — *If low: Chapters 20–22*

Your Profile

Dimension	Score (5-25)	Status
1. Awareness	___	__________
2. Strategic	___	__________
3. Cultural	___	__________
4. Leadership	___	__________
5. Operational	___	__________
6. Data & Technology	___	__________
7. Measurement	___	__________
8. Implementation	___	__________
9. Sustainability	___	__________

Status key: 21–25 Strong · 16–20 Developing · 11–15 At Risk · 5–10 Not Ready

Reading your profile: Don't start with your lowest score — start with the dimension that creates the greatest downstream impact. Strategic and Leadership Readiness are load-bearing; Measurement Readiness is the dimension that tells you whether the others are improving. Fix upstream first.

Re-score: quarterly while building the capability; twice yearly once mature; always before a new initiative and after any major organizational change — especially a merger or acquisition.

The AI Ready Scorecard™ is part of the AI Ready System™ by Valetta Taylor. Full-size worksheets, team scoring templates, and the companion assessment: ecfsystem.com/aiready.

NOTES

All claims in Chapter 3 re-verified against original sources. Formatted for the book's endnotes section; re-check once more at final proof stage.

Claim 1: "Nearly four in ten U.S. workers worry AI may make some or all of their job duties obsolete."

→ American Psychological Association, *2023 Work in America Survey: Artificial Intelligence, Monitoring Technology, and Psychological Well-Being* (conducted by The Harris Poll, April 17–27, 2023; n = 2,515 employed adults): 38%. The 2025 Work in America wave reports the same figure (38%). apa.org/pubs/reports/work-in-america/2023-work-america-ai-monitoring

Claim 2: "Among workers worried about AI, roughly half said their work negatively impacts their mental health — compared to fewer than a third of workers without that worry."

→ Same APA 2023 survey: 51% vs. 29%.

Claim 3: "Nearly two-thirds of AI-worried workers reported feeling tense or stressed during a typical workday, versus just over a third of everyone else."

→ Same APA 2023 survey: 64% vs. 38%. (The 2024 wave found 54% vs. 36% — directionally identical.)

Claim 4: "Pew Research Center found over half of U.S. workers worried about AI's future impact in the workplace, with the worried outnumbering the hopeful."

→ Pew Research Center, *U.S. Workers Are More Worried Than Hopeful About Future AI Use in the Workplace* (survey Oct. 7–13, 2024; n = 5,273 employed U.S. adults; published Feb. 25, 2025): 52% worried vs. 36% hopeful; 33% overwhelmed. pewresearch.org/social-trends/2025/02/25/

Claim 5: "Mercer's global survey of twelve thousand workers found the share who fear losing their job to AI climbing sharply in just two years."

→ Mercer, *Global Talent Trends 2026* (12,000 workers and business leaders worldwide): 40% fear AI will make their job obsolete, up from 28% in 2024 — a 12-point rise in two years.

Claim 6: "APA's research found middle managers and frontline workers among the most likely to fear their duties becoming obsolete."

→ APA, *2025 Work in America Survey*: approximately 40% of frontline workers and middle managers expressed this concern, versus roughly 29% of upper managers.

Claim 7: "Workers worried about AI were significantly more likely to say they intended to look for a new job within the year."

→ APA 2023: 46% of AI-worried workers intended to look for another job vs. 25% of the unworried. (2024 wave: 36% vs. lower among unworried — direction holds across waves.)

Claim 8: "The APA's own guidance for employers is that transparent, honest communication about AI — educating people, involving them, giving them a voice — measurably reduces the stress."

→ APA 2023 Work in America report, employer recommendations, and accompanying APA commentary (apa.org/topics/healthy-workplaces/artificial-intelligence-workplace-worry) emphasizing communication, education, and employee involvement as the response to AI-related worry.

Claim 9 (added July 2026): "Research out of Harvard Business Review found that AI adoption is quietly overloading middle managers — the layer expected to implement the technology, maintain quality, and develop their teams through the change, all at once."

→ Julia Shin and Sandra J. Sucher, "AI Adoption Is Overloading Your Middle Managers," *Harvard Business Review*, June 2026. hbr.org/2026/06/ai-adoption-is-overloading-your-middle-managers

Production note: at final proof, refresh Claims 4–6 in particular — Pew and APA field new waves annually, and using the most current wave at publication keeps the book's

shelf life longest. All current figures support the chapter's language without edits.

About the Author

Valetta Taylor is a transformation executive with more than twenty years in operations and transformation, including over a decade embedded in private equity-backed organizations. She has led ten to fifteen post-acquisition integrations end to end across waste and recycling, aviation, SaaS, healthcare, and customer operations — the kind of work where a plan meets an organization and one of them gives way.

She is the creator of the Executive Conditions Framework™, a diagnostic model that identifies the structural conditions determining whether transformation holds: strategic clarity, decision velocity, operational alignment, leadership ownership, and cultural safety. The framework grew out of a simple and repeatedly confirmed observation — tools don't transform organizations. Conditions do.

That observation is why this book exists. Artificial intelligence is the most capable set of tools executives have ever been handed, and capability has never been the thing that determines whether a technology takes hold. Valetta wrote AI Ready in 90 Days for leaders who have watched a well-funded initiative stall for reasons no one could name, and who would rather diagnose the ground before planting in it again.

She is the author of EXPOSED: When the Conditions Aren't Right, and the founder of The Harmony Collection by Valetta Taylor, LLC. She lives and works in Houston, Texas.

ecfsystem.com

www.ingramcontent.com/pod-product-compliance
Lightning Source LLC
LaVergne TN
LVHW100729061126
851857LV00011B/460
* 9 7 9 8 9 9 6 5 9 0 1 2 4 *